Lecture Notes in Mathematics

Edited by A. Dold and B. Eckmann

Series: Mathematisches Institut der Univ

Adviser: F. Hirzebruch

439

Kenji Ueno

Classification Theory of Algebraic Varieties and Compact Complex Spaces

Notes written in collaboration with P. Cherenack

Springer-Verlag
Berlin · Heidelberg · New York 1975

Dr. Kenji Ueno
Department of Mathematics
Faculty of Science
University of Tokyo
Tokyo/Japan

Library of Congress Cataloging in Publication Data

Ueno, Kenji, 1945-
Classification theory of algebraic varieties and compact complex spaces.

(Lecture notes in mathematics ; 439)
Bibliography: p.
Includes index.
1. Algebraic varieties. 2. Complex manifolds.
3. Analytic spaces. 4. Fiber bundles (Mathematics)
I. Title. II. Series: Lecture notes in mathematics (Berlin) ; 439.
QA3.L28 no. 439 [QA564] 510'.8s [514'.224] 75-1211

AMS Subject Classifications (1970): 14-02, 14A10, 14J15, 32-02, 32C10, 32J15, 32J99, 32L05

ISBN 3-540-07138-5 Springer-Verlag Berlin · Heidelberg · New York
ISBN 0-387-07138-5 Springer-Verlag New York · Heidelberg · Berlin

Offsetdruck: Julius Beltz, Hemsbach/Bergstr.

To Professor K. Kodaira

PREFACE

The present notes are based on the lectures which I gave at the University of Mannheim from March 1972 to July 1972. The lectures were informal and were intended to provide an introduction to the classification theory of higher dimensional algebraic varieties and compact complex spaces recently developed by S. Kawai, S. Iitaka and other mathematicians in Tokyo. The notes were taken by P. Cherenack. Since there were no available lecture notes on these subjects, I decided after reading Cherenack's notes to rewrite them more systematically so that they would serve as an introduction to our classification theory. Several topics which I did not mention at Mannheim have been added.

P. Cherenack typed a good part of the first version of my manuscript, improving my English. He also compiled a first version of the bibliography which was quite helpful in completing the final version of the bibliography. Here I gratefully acknowledge my indebtedness to him.

I would like to express my thanks to Professor H. Popp and the Department of Mathematics of the University of Mannheim for giving me the opportunity of visiting Mannheim and of giving these lectures. The greater part of the final version of the present notes was written when I was a visiting member of the Mathematical Institute of the University of Bonn. I wish to express my thanks to Professor F. Hirzebruch and the Mathematical Institute of the University of Bonn for inviting me to Bonn, and to the Department of Mathematics of the University of Tokyo for giving me permission to visit Mannheim and Bonn.

My thanks are due to Dr. Y. Namikawa, Dr. E. Horikawa, Mr. T. Fujita and Mr. Masahide Kato who read the manuscript in whole or in part and pointed out some mistakes and suggested some improvements. I wish to express my thanks to Professor H. Popp and Professor S. Iitaka for their constant encouragement during the preparation of the present notes. Last, but not least, I would like to express my thanks to Miss K. Motoishi for her typewriting.

Tokyo, September 1974. Kenji Ueno.

Table of Contents

INTRODUCTION

There are two notions of classification theory of complete algebraic manifolds defined over $\mathbb{C}$ or compact complex manifolds, a rough classification and a fine classification. For example, in the case of non-singular curves, i.e. compact Riemann surfaces, we subdivide isomorphism classes of curves into infinitely many families M_g, $g=1, 2, 3, \ldots$, where M_g consists of isomorphism classes of curves of genus g. This gives a rough classification. The study of the structure of the set M_g gives a fine classification of curves. This is usually called the theory of moduli. The study of the set M_g is nothing but the study of all complex structures on a fixed topological model of a compact Riemann surface of genus g. One of the important results on the theory of moduli of curves is that M_g carries the structure of a quasi-projective variety (see Baily [1] and Mumford [5]).

In these lecture notes we shall mainly discuss a rough classification of birational (resp. bimeromorphic) equivalence classes of complete algebraic manifolds (resp. compact complex manifolds). But we shall show later that the theory of moduli, that is the finest classification, is deeply related to a rough classification of compact complex manifolds without non-constant meromorphic functions (see §13 below). On the other hand, since we are interested in birational (resp. bimeromorphic) equivalence classes of algebraic manifolds (resp. compact complex manifolds), by virtue of the resolution of singularities due to Hironaka ([1], [2]), we can consider classification of algebraic varieties and reduced, irreducible, compact complex spaces.

Thus one of the main purposes of our classification theory is to find good birational (resp. bimeromorphic) invariants of algebraic manifolds (resp. of compact complex manifolds), such as the genera of curves so that we can subdivide birational (resp. bimeromorphic) equivalence classes of manifolds into certain large families.

Let us recall briefly the classification theory of analytic surfaces, i.e. two-dimensional compact complex manifolds. The classification of surfaces is much more complicated than the classification of curves. Contrary to the case of curves, two bimeromorphically equivalent surfaces are not necessarily analytically isomorphic to each other. The difficulty is overcome using the theory of exceptional curves of the first kind (see Definition 20.1 below and Zariski [4]). The theory says that any analytic surface is obtained from a relatively minimal model by finite succession of monoidal transformations. Moreover, except for rational and ruled surfaces, two relatively minimal models are bimeromorphically equivalent if and only if they are isomorphic to each other (see Theorem 20. 3). Thus it is enough to classify relatively minimal models of surfaces.

Classification of algebraic surfaces was done partly by Castelnuovo and mainly by Enriques. They found the important birational invariants, the irregularity and the plurigenera of a surface. The irregularity $q(S)$ of a surface S is defined by

$$q(S) = \dim_{\mathbf{C}} H^1(S, \underline{O}_S).$$

For a positive integer m, the m-genus $P_m(S)$ of a surface S (if we do not specify the integer m, we call it a plurigenus of S) is defined by

$$P_m(S) = \dim_{\mathbf{C}} H^0(S, \underline{O}(mK_S)),$$

where K_S is the canonical bundle of the surface S. (Note that the above definitions work for any compact complex manifold.) According to whether $P_{12} = 0$, $P_{12} = 1$ or $P_{12} > 1$, respectively, they classified algebraic surfaces into three big classes (using the Kodaira dimension κ which will be defined in §6 below, we can say that these three classes consist respectively of the algebraic surfaces for which $\kappa = -\infty$, $\kappa = 0$, $\kappa > 0$), and subdivided each class into finer classes (see Enriques [1], p.463-464). Their arguments were quite intuitive. Rigorous proofs were given by several mathematicians, especially by Kodaira (see Kodaira [2], [3], Šafarevič et al [1] and Zariski [7], II, p.277-505).

Kodaira has generalized the classification theory of algebraic surfaces to that of analytic surfaces (see Kodaira [2], [3]). In Kodaira [2], I, we can find two important results on non-algebraic analytic surfaces. The one is the algebraic reduction of analytic surfaces and the other is the structure theorem of Kähler surfaces without non-constant meromorphic functions. Let $\mathbb{C}(S)$ be the field consisting of all meromorphic functions on an analytic surface S. The transcendence degree over $\mathbb{C}$ of the function field $\mathbb{C}(S)$ is called the algebraic dimension $a(S)$ of S. Kodaira has shown that if $a(S) = 1$, there exist a non-singular curve C and a subjective morphism $\varphi : S \longrightarrow C$, which we call the algebraic reduction of S, such that φ induces an isomorphism between the function fields $\mathbb{C}(C)$ and $\mathbb{C}(S)$, and that general fibres of the morphism φ are elliptic curves. Kodaira has also shown that if a Kähler surface S is of algebraic dimension zero, then the irregularity $q(S)$ of S is zero or two, and that if moreover $q(S) = 2$ the natural mapping of S into its Albanese variety is a modification of S. These two results have been generalized by Kawai in the case of three-dimensional compact complex manifolds (see Kawai [1] and §12, §13 below). Kawai's results were the first definite results on classification theory of higher dimensional compact complex manifolds.

By the way, we have already mentioned that the pulrigenera play an important role in the classification of surfaces. The plurigenera are deeply related to the pluricanonical mappings. Let us suppose that $P_m(S) \neq 0$ for an analytic surface S and a positive integer m. Let $\{\varphi_0, \varphi_1, \cdots, \varphi_N\}$ be a basis of the vector space $H^0(S, \underline{O}(mK_S))$. We define a meromorphic mapping $\Phi_{mK} : S \longrightarrow \mathbb{P}^N$ by

$$\begin{array}{ccc} S & \longrightarrow & \mathbb{P}^N \\ \cup\!\!\!\shortmid & & \cup\!\!\!\shortmid \\ z & \longrightarrow & (\varphi_0(z): \varphi_1(z): \cdots : \varphi_N(z)) \end{array}$$

and call it the m-th canonical mapping. If the integer m is not specified, we call it a pluricanonical mapping. Enriques had already studied the pluricanonical mappings of certain algebraic surfaces of general type (see Enriques [1]) and Kodaira has given a general theory of the pluricanonical mappings of algebraic surfaces of general type. The nature of the pluricanonical mappings of elliptic surfaces can be easily deduced from the canonical bundle formula for elliptic surfaces due to Kodaira (see Kodaira [3], I and the formula 20.13.1 below).

Inspired by these results, Iitaka has studied the pluricanonical mappings of higher dimensional compact complex manifolds (see Iitaka [2] and §5 − §8 below). More generally he has studied meromorphic mappings associated with complete linear systems of Cartier divisors on a normal variety. He has defined the Kodaira dimensions of compact complex manifolds and has proved a fundamental theorem on the pluricanonical fibrations (see Theorem 6.11 below). The Kodaira dimension

$\kappa(M)$ of a compact complex manifold M is, by definition, $-\infty$ if $P_m(M) = 0$ for every positive integer, and is the maximal dimension of the image varieties of M under the pluricanonical mappings if $P_m(M) \geq 1$ for at least one positive integer m. The Kodaira dimension is a bimeromorphic invariant of a given compact complex manifold. The fundamental theorem on the pluricanonical fibrations due to Iitaka says that if the Kodaira dimension $\kappa(M)$ of a compact complex manifold M is positive, then there exists a bimeromorphically equivalent model M^* of M which has the structure of a fibre space whose general fibres are of Kodaira dimension zero (for the precise statement of the theorem, see Theorem 6.11).

As we have mentioned above, with each of the bimeromorphic invariants, that is, with the algebraic dimension, with the Kodaira dimension, with each of the plurigenera, and with the irregularity we can associate a meromorphic mapping and introduce a fibre space structure on the compact complex manifold. With the algebraic dimension, we can associate an algebraic reduction. With the Kodaira dimension and with each plurigenus we can associate a pluricanonical mapping. With the irregularity (if a complex manifold is neither algebraic nor Kähler, we should replace the irregularity by the Albanese dimension), we can associate the Albanese mapping. Using the fibre spaces introduced by these mappings, we shall show that the classification theory is reduced to the study of these fibre spaces and the study of special manifolds.

In Iitaka [3], relying on the fundamental theorem of the pluricanonical fibrations and Kawai's results mentioned above, Iitaka has discussed the classification theory of algebraic varieties and compact complex spaces. After his paper a number of interesting results on classification theory have been obtained. Nakamura has shown that the Kodaira dimension is not necessarily invariant under small deformations (see Nakamura [1] and § 17 below). Nakamura and Ueno have shown the addition formula for Kodaira dimensions of analytic fibre bundles whose fibres are Moishezon manifolds (see Nakamura and Ueno [1], and § 15 below). This formula gives an affirmative answer to a special case of Conjecture C_n (see § 11). Ueno has studied Albanese mappings and has shown that Albanese mappings play an important role in classification theory (see Ueno [3] and Chapter IV below). In Ueno [3], the author has also proved the canonical bundle formula for certain elliptic three-folds and has studied Kummer manifolds (see §11 and §16 below). Akao has studied prehomogeneous Kähler manifolds (see Akao [2] and §19). Kato has studied complex structures on $S^1 \times S^5$ (see Masahide Kato [2], [3] and § 18). Iitaka has introduced new birational invariants and has studied three-dimensional rational manifolds (see Iitaka [4] and § 19). He has also studied three-dimensional algebraic manifold whose universal covering is the three-dimensional complex affine space (see Iitaka [5]).

The main purpose of the present lecture notes is to provide a systematic treatment of these subjects. Many examples of complex manifolds exhibiting a difference between the classification of surfaces and that of higher dimensional complex manifolds will be given (see Chapter VII below). At the moment our classification theory is far from satisfactory but we already have a lot of interesting results. We hope that the present notes will serve as an introduction to this new field.

For a complete understanding of these lecture notes, a knowledge of the general theory of complex manifolds and of the classification theory of surfaces is indispensable. On these subjects we refer the reader to Kodaira and Morrow [1] and to Kodaira [2], [3]. Ueno [4], [5] will serve as an introduction to the present notes.

Conventions and Notations

Unless otherwise explicitly mentioned, the following conventions will be in force throughout these notes.

1) By an algebraic variety we mean a complete irreducible algebraic variety defined over $\mathbb{C}$. By an algebraic manifold we mean a non-singular algebraic variety.

2) All complex manifolds are assumed to be compact and connected.

3) By a complex variety we mean a compact irreducible reduced complex space.

4) The word "manifold"(resp. "variety") means an algebraic manifold (resp. algebraic variety) or a complex manifold (resp. complex variety).

5) For a Cartier divisor D on a variety D, by [D] we denote the complex line bundle associated with the divisor D. We often identify the sheaf $\underline{O}_V(mD)$ with the sheaf $\underline{O}([mD])$ by a natural isomorphism. In §7 we shall distinguish these two sheaves.

6) For a complex line bundle L we often write mL instead of $L^{\otimes m}$.

7) By an elliptic bundle over a manifold M we mean a fibre bundle over M whose fibre and structure group are an elliptic curve E and the automorphism group Aut(E) of E, respectively.

8) The dimension of a variety is a complex dimension.

9) By a fibre space $f : M \longrightarrow W$ of complex manifolds we mean that a morphism f is surjective and that any fibre of f is connected.

10) For a Cartier divisor D on a complex variety V, we define $\ell(mD)$ by

$$\ell(mD) = \dim_{\mathbb{C}} H^0(V^*, \underline{O}_{V^*}(m\iota^*D))$$

where $\iota : V^* \longrightarrow V$ is the normalization of V.

$a(V)$;	the algebraic dimensions of a complex variety V (see Definition 3.2).
$(A(M), \alpha)$;	the Albanese torus of a complex manifold M (see Definition 9.6).
$\mathrm{Aut}(V)$;	the automorphism group of a complex variety V.
$\mathrm{Aut}^0(V)$;	the identity component of $\mathrm{Aut}(V)$.
$\mathrm{Bim}(V)$;	the bimeromorphic transformation group of a complex variety V.
$\mathbb{C}(V)$;	the meromorphic function field of a variety V.
$g_k(V)$;	see Definition 9.20.
$K_M = K(M)$;	the canonical line bundle (a canonical divisor) of a manifold M.
$\mathbb{P}(\underline{F})$;	the projective fibre space associated with a coherent sheaf $\underline{F}$ (see 2.8).
$p_g(V)$;	the geometric genus of a variety V (see Definition 6.5).
$P_m(V)$;	the m-genus of a variety V (see Definition 6.5).
$q(V)$;	the irregularity of a variety V (see Definition 9.20).
$q_k(V)$;	the k-th irregularity of a variety V (see Definition 9.20).
$r(V)$;	see Definition 9.20.
$t(V)$;	the Albanese dimension of a variety V (see Definition 9.21).
$\alpha : M \longrightarrow A(M)$;	the Albanese mapping (see Definition 9.6).
$\kappa(D, V)$;	the D-dimension of a variety V (see Definition 5.1).
$\kappa(V)$;	the Kodaira dimension of a variety V (see Definition 6.5).
$\Phi_{mD} : V \longrightarrow \mathbb{P}^N$;	the meromorphic mapping associated with a complete linear system $\|mD\|$ (see 2.4 and §5). If D is the canonical line bundle, we call it the m-th canonical mapping.

Σ^k ; a real k-dimensional homotopy sphere which bounds a parallelizable manifold.

Ω^k_M : the sheaf of germs of holomorphic k-forms on a complex manifold M.

Chapter I

Analytic spaces and algebraic varieties

For the reader, in this chapter, we collect some important results on analytic spaces and algebraic varieties. Almost all these results shall be given without proof. References will be found in the text. Furthermore, we do not necessarily provide any of the notions from the function theory of several complex variables required for a complete understanding of this book. The reader could consult Grauert and Remmert [1], Gunning and Rossi [1], Hitotsumatsu [1], Hörmander [1] and Narasimhan [1].

In §1, at the beginning, the comparison theorem of algebraic geometry and analytic geometry (usually quoted as GAGA) can be found. In this book we shall use complex analytic methods to study algebraic varieties. However, sometimes, it is useful to use algebraic methods. GAGA assures us that both methods yield the same results under suitable conditions. Next, the Grauert proper mapping theorem, the existence theorem of the Stein factorization and Zariski's Main Theorem are introduced. These theorems play a fundamental role in our classification theorey.

In §2, we shall first give some results on meromorphic mappings. Here we employ the definition of the meromorphic mapping due to Remmert. The meromorphic mapping Φ_L associated with a line bundle L on a complex variety (see Example 2.4.2, below) will be studied in detail in Chapter II and Chapter III. For that purpose, it is convenient to introduce the notion of the "projective fibre space $\mathbb{P}(\underline{F})$ associated with a coherent sheaf $\underline{F}$ on a complex variety" due to Grothendieck. Finally, there are some important results on resolutions of singularities due to Hironaka.

In § 3, the algebraic dimension of a complex variety and an algebraic reduction of a complex variety will be defined. The algebraic dimension is the most fundamental and important bimeromorphic invariant of a complex variety. If the algebraic dimension $a(M)$ of a complex variety M is smaller than $\dim M$, by an algebraic reduction of M, a certain bimeromorphically equivalent variety M^* of M has the structure of a fibre space over a projective manifold. Such fibre spaces will be studied in detail in Chapter V, § 12.

§1. GAGA, Proper mapping theorem, Stein factorization

In this section, algebraic varieties may not be complete and complex spaces may not be reduced, unless explicitly otherwise mentioned.

Let $(Sch/\mathbb{C})$ be the category of schemes of finite type over $\mathbb{C}$ (we often call them algebraic $\mathbb{C}$-schemes) and let (An) be the category of analytic spaces.

It is well known that a complex algebraic varieties can be canonically identified with an irreducible reduced algebraic $\mathbb{C}$-scheme (see for example, Mumford [1], Chap II, §3).

Let $(V, \underline{O}_V)$ be an algebraic $\mathbb{C}$-scheme and let $\{V_i\}_{i\in I}$ be an affine open covering of V. $V_i = \mathrm{Spec}\ R_i$, $R_i = \mathbb{C}[X_{i1}, X_{i2}, \ldots, X_{in_i}]/\underline{J}_i$. Then, the ideal $\underline{J}_i$ defines a closed analytic subspace V_i^{an} of $\mathbb{C}^{n_i}$. Patching together all V_i^{an}'s we obtain a complex space $(V^{an}, \underline{O}_{V^{an}})$. It is easy to see that the analytic space $(V^{an}, \underline{O}_{V^{an}})$ depends only on the scheme $(V, \underline{O}_V)$, but does not depend on a choice of affine coverings. Moreover, for a morphism $f : V \longrightarrow W$ of algebraic $\mathbb{C}$-schemes, there is a morphism $f^{an} : V^{an} \longrightarrow W^{an}$ of associated complex spaces. Thus we have a functor $\psi : (Sch/\mathbb{C}) \longrightarrow (An)$ (Serre [3]). Notice that the topology of the underlying topological space of $(V, \underline{O}_V)$ (the Zariski topology) is weaker than that of $(V^{an}, \underline{O}_{V^{an}})$ (the complex topology) if $\dim V \geqq 1$.

Consider the structure sheaves $\underline{O}_V$ and $\underline{O}_{V^{an}}$ on V and V^{an}, respectively. (Coh/V) will denote the category whose objects are the algebraic coherent sheaves over V and whose morphisms are the $\underline{O}_V$-linear homomorphisms of sheaves over V. Similarly, (Coh/V^{an}) denotes the category whose objects are the analytic coherent sheaves over the analytic space V^{an} associated with V and whose morphisms are the $O_{\underline{V}^{an}}$-linear homomorphisms of sheaves over V^{an}. Then the functor ψ induces a functor

$$\psi \ : \ (Coh/V) \longrightarrow (Coh/V^{an}) \ .$$

If $\underline{F}$ is an object of (Coh/V), we write $\psi(\underline{F}) = \underline{F}^{an}$.

We now describe some of the properties which the functor ψ preserves and reflects, in the following two lemmas.

<u>Lemma 1.1</u>. Let V be a scheme of finite type over $\mathbb{C}$. V is smooth, normal, reduced, proper, connected or irreducible if and only if the associated complex space V^{an} is smooth, normal, reduced, compact, connected or irreducible, respectively.

<u>Lemma 1.2</u>. Let $f : V \longrightarrow W$ be a morphism of schemes of finite type over $\mathbb{C}$. The morphism f is flat, unramified, etale, smooth, an isomorphism, an open immersion, a closed immersion, surjective or proper if and only if the associated morphism $\psi(f) = f^{an}$ has the same property.

If $\underline{C}$ is a category and b is an object of $\underline{C}$, then $\underline{C}/b$ will denote the category whose object are the arrows $a \longrightarrow b$ in $\underline{C}$ and whose morphisms are the commutative diagrams

$$\begin{array}{ccc} a & \longrightarrow & a' \\ \downarrow & & \downarrow \\ b & = & b \end{array}$$

in $\underline{C}$. Let $Fin(Sch/V)$ be the full subcategory of (Sch/V) whose objects are schemes finite over V. $Et(Sch/V)$ is the full subcategory of $Fin(Sch/V)$ consisting of those schemes finite and etale over V. In a similar way, we define and designate elements of $Fin(An/V^{an})$ and $Et(An/V^{an})$.

<u>Theorem 1.3</u>. (GAGA). 1) Prop(Sch/$\mathbb{C}$) denotes the full subcategory of (Sch/$\mathbb{C}$) whose objects are proper schemes over $\mathbb{C}$. The functor ψ restricted to Prop(Sch/$\mathbb{C}$),

$$\hat{\psi} : \text{Prop}(\text{Sch}/\mathbb{C}) \longrightarrow (\text{An}),$$

is fully faithful.

2) Let V be a proper $\mathbb{C}$-scheme. Then the functor ψ induces an equivalence of categories

$$\tilde{\psi} : \text{Fin}(\text{Sch}/V) \longrightarrow \text{Fin}(\text{An}/V)$$

which restricts to another equivalence of categories

$$\bar{\psi} : \text{Et}(\text{Sch}/V) \longrightarrow \text{Et}(\text{An}/V).$$

3) Suppose that V is a proper $\mathbb{C}$-scheme.
The functor $\psi : (\text{Coh}/V) \longrightarrow (\text{Coh}/V^{an})$ is an equivalence of categories. It follows, in this case, that there is a canonical isomorphism

$$H^p(V, \underline{F}) \xrightarrow{\sim} H^p(V^{an}, \underline{F}^{an}),$$

where $\underline{F}$ is an element of (Coh/V) and p is a non-negative integer. Moreover, by ψ, the category of algebraic coherent locally free sheaves and the category of analytic coherent locally free sheaves are equivalent.

4) More generally, let $f : V \longrightarrow W$ be a proper morphism in (Sch/$\mathbb{C}$). There exists a canonical isomorphism

$$\theta_p : (R^p f_*(\underline{F}))^{an} \xrightarrow{\sim} R^p f_*^{an}(\underline{F}^{an})$$

for any algebraic coherent sheaf $\underline{F}$ on V and for any non-negative integer p.

For the proofs of Lemma 1.1, Lemma 1.2 and Theorem 1.3 and, in addition, for details about these results, the reader should consult Serre [3] and SGA 1, Exp. XII.

In view of Lemma 1.1, Lemma 1.2 and Theorem 1.3 , if V is an algebraic variety, we never distinguish, unless explicitly stated otherwise, between the algebraic structure V and the associated analytic structure V^{an}. If we must consider V as the $\mathbb{C}$-scheme or the algebraic variety, we often use the notation V^s or V^{alg}.

We state now the proper mapping theorem.

Theorem 1.4. (Grauert). If $f : X \longrightarrow Y$ is a proper morphism of analytic spaces and $\underline{F}$ is a coherent analytic sheaf on X, then we have :

1) The p-th direct image sheaf $R^p f_* \underline{F}$ is coherent for any non-negative integer p.

Suppose moreover that the sheaf $\underline{F}$ is flat over Y. Then we have:

2) $d_p(y)=\dim_{\mathbb{C}} H^p(X_y, \underline{F}_y)$ is an upper semi-continuous function on the underlying topological space $|Y|$ of Y. Here $X_y = f^{-1}(y)$ and $\underline{F}_y$ is the restriction of the sheaf $\underline{F}$ to X_y.

3) Suppose that Y is reduced and connected. If the above function $d_p(y)$ is a constant m on $|Y|$, then $R^p f_* \underline{F}$ is a locally free sheaf of rank m. In this case a natural homomorphism

$$\lambda_y : (R^p f_* \underline{F})_y \underset{\underline{O}_{Y,y}}{\otimes} \mathbb{C} \longrightarrow H^p(X_y, \underline{F}_y)$$

is an isomorphism for any $y \in Y$. In general λ_y is an isomorphism for any point y outside a nowhere dense analytic subset of Y.

For a proof of this theorem, see Grauert [1], Kiehl and Verdier [1], Knorr [1], Forster and Knorr [1] and Riemenschneider [1].

Remark 1.5. 1). Suppose that $f : V \longrightarrow W$ is a proper morphism of algebraic $\mathbb{C}$-schemes. We find, by GAGA (Theorem 1.3), that Theorem 1.4 is true if $\underline{F}$ is an algebraic coherent sheaf. The algebraic version of Theorem 1.4 has been proved in EGA III in an easier manner, however.

2) Let $f : X \longrightarrow Y$ be a proper morphism of analytic spaces and let $\underline{F}$ be a coherent analytic sheaf on X. If Y is reduced, there exists an open dense subset U of Y such that $\underline{F}$ is flat over U (see Frisch [1], Kiehl [1]). Hence for our later application, the restriction that $\underline{F}$ is flat on Y in the above Theorem 1.3, 2), 3) is not essential.

Corollary 1.6. Consider a proper morphism $f : X \longrightarrow Y$ of analytic spaces and suppose that S is an analytic subset of X. Then the image $f(S)$ of S is an analytic subset of Y.

Proof. Let $\underline{J}$ be the sheaf of ideals of definition of S in X. As $\underline{J}$ is coherent, $\underline{O}_X/\underline{J}$ is coherent. By the theorem, $\underline{F} = f_*(\underline{O}_X/\underline{J})$ is coherent. We have $f(S) = \{y \in Y | \underline{F}_y \neq 0\}$. From the coherency of $\underline{F}$, it follows that $f(S)$ is an analytic subset of Y. Q.E.D.

Corollary 1.7. If $f : X \longrightarrow Y$ is a proper surjective morphism of smooth analytic spaces, i.e. complex manifolds, there exists a nowhere dense analytic subset S of X such that f is smooth (=of maximal rank) at any point of $X - S$ and $f(S)$ is a nowhere dense analytic subset of Y.

Proof. $S = \{x \in X \mid (df)_x \text{ is not of maximal rank}\}$ is an analytic subset of X. The previous corollary implies that $f(S)$ is an analytic subset of Y. Finally, the following theorem of Sard applies and shows that $f(S)$ is a nowhere dense subset of Y. Q.E.D.

Theorem (Sard). When $f : M \longrightarrow N$ is a differentiable map of differntiable manifolds, and S is the set of all points in M

where $(df)_x$ is not of maximal rank, $f(S)$ has measure zero in N.

A proof of this theroem can be found in Matsushima [1].

Corollary 1.8. Let $f : X \longrightarrow Y$ be a proper surjective morphism of complex manifolds. There is a nowhere dense analytic subset S' of Y such that $y \notin S'$ implies that $X_y = f^{-1}(y)$ is smooth, i.e., the general fibre of F is smooth.

Proof. Let S be the analytic subset defined in the proof of Corollary 1.7. Then, by Corollary 1.7, $S' = f(S)$ is a nowhere dense analytic subset of Y. Suppose that $y \notin S'$ and $f(x) = y$. There are coordinate neighbourhoods U and V of x and y, respectively, where f is defined by

$$z^i = f^i(w^1, \dots, w^n),$$

$i = 1, \dots, m$. Here the local coordinates on V(resp. U) are $z^1, \dots, z^m$ (resp. $w^1, \dots, w^n$). But, by a well-known criterium for non-singularity, x is a smooth point of the analytic set $f^{-1}(y)$ if and only if

$$(df)_x = \left(\frac{\partial f^i}{\partial w^j}\right)_x$$

attains its maximal rank at x. Hence Corollary 1.8 follows from Corollary 1.7. Q.E.D.

A reduced, irreducible, compact analytic space will be called a complex variety. The next theorem can often be applied to determine when the fibres of a map between complex varieties is connected.

Theorem 1.9. Let $f : X \longrightarrow Y$ be a proper surjective morphism of reduced complex spaces. There is a commutative diagram, called the Stein factorization of f,

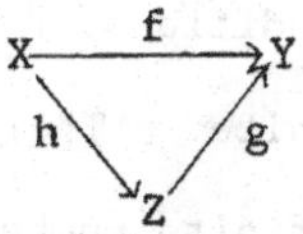

in the category (An) where we have the following properties.

1) Z is a reduced complex space, and h and g are surjective.

2) The fibres of h are connected.

3) The fibres of g consist of a finite number of points. That is, g is a finite ramified covering.

4) If $y \in Y$, the points of $g^{-1}(y)$ are in bijective correspondence with the connected components of $f^{-1}(y)$.

The proof is found in Cartan [2].

<u>Corollary 1.10</u>. Suppose that $f : X \longrightarrow Y$ is a proper surjective morphism of reduced complex spaces. Moreover assume that Y is a normal algebraic variety and the <u>rational</u> function field $\mathbb{C}(Y)$ of Y is algebrically closed in the meromorphic function field $\mathbb{C}(X)$ of X via f. Then any fibre of f is connected.

<u>Proof</u>. Let

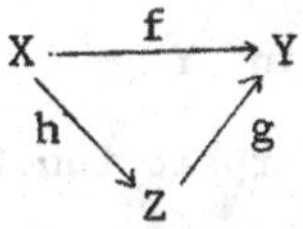

be the Stein factorization of $f : X \longrightarrow Y$. By our assumption $\mathbb{C}(Y)$ and $\mathbb{C}(Z)$ are isomorphic via g. Hence $g : Z \longrightarrow Y$ is a finite birational morphism. As Y is normal, by Zariski's Main Theorem below, g must be an isomorphism. Q.E.D.

<u>Theorem 1.11</u>. (Zariski's Main Theorem). Let $f : X \longrightarrow Y$ be a proper surjective morphism of irreducible reduced complex spaces

which satisfies the following conditions.

1) The morphism f is finite. That is, for any point $y \in Y$, $f^{-1}(y)$ consists of a finite number of points.

2) There exists an open set U of Y such that, for any point $u \in U$, $f^{-1}(u)$ is one point.

3) Y is a normal complex space.

Then the morphism f is an isomorphism.

Proof. We set $\underline{F} = f_* \underline{O}_X$. $\underline{F}$ is coherent and a finite $\underline{O}_Y$ algebra. By Sem. Cartan 60/61, Exp. 19, Théorème 2, $f : X \longrightarrow Y$ is analytically isomorphic to Specan$(\underline{F}) \longrightarrow Y$. There is a nowhere dense analytic subset S of Y such that $\underline{F}$ is locally free on $Y - S$. On the other hand, as Y is normal, from the condition 2) we infer readily that $f_* \underline{O}_{X|U} = \underline{O}_U$. (Note that, for any open set V in U, a holomorphic function h on $f^{-1}(V)$ induces a continuous function $\hat{h}$ on V, which is holomorphic by virtue of normality of V.) Hence on $Y - S$, $\underline{F}$ is an invertible sheaf and $X - f^{-1}(S)$ is isomorphic to $Y - S$. Now we shall prove that S is empty.

Let g be a bounded holomorphic function on $f^{-1}(V)$ where V is an open neighbourhood of $x \in S$ in Y. As $f^{-1}(V - S)$ and $V - S$ are isomorphic, there exists a holomorphic function h' on $V - S$ such that $g = h' \circ f$ on $f^{-1}(V - S)$. As $V \cap S$ is an analytic subset of V, h' is bounded on $V - S$ and V is normal, the holomorphic function h' can be extended to a holomorphic function g on V (see for example, Grauert and Remmert [1], Satz 22, p.286). Then we have $g = h \circ f$. This implies that

$$\underline{F}_x = \underline{O}_{Y,x} .$$

Hence S must be empty. Q.E.D.

<u>Corollary 1.12</u>. Let $f : X \longrightarrow Y$ be a proper surjective morphism of connected reduced complex spaces such that

1) Y is normal ;

2) there exists an open set U of Y such that for any $y \in U$, the fibre $X_y = f^{-1}(y)$ is connected.

Then all the fibres of f are connected.

<u>Proof</u>. Let

$$\begin{array}{ccc} X & \xrightarrow{\quad f \quad} & Y \\ & {\scriptstyle g}\searrow \quad \nearrow{\scriptstyle h} & \\ & W & \end{array}$$

be the Stein factorization of f. Then, by the conditions 1), 2), above, the morphism $h : W \longrightarrow Y$ satisfies the conditions 1), 2), 3) in Theorem 1.11. Hence h is an isomorphism. Q.E.D.

<u>Proposition 1.13</u>. Let $f : X \longrightarrow Y$ be a proper surjective morphism of reduced complex spaces such that Y is normal and every fibre of f is connected. Then

$$f_* \underline{O}_X = \underline{O}_Y .$$

<u>Proof</u>. We shall show that $(f_* \underline{O}_X)_y = \underline{O}_{Y,y}$ for any $y \in Y$. Let U be an open neighbourhood of y in Y and let g be a holomorphic function on $f^{-1}(U)$. We can choose U so small that g is bounded on $f^{-1}(U)$. There exists a nowhere dense analytic subset S in U such that for any point $u \in U' = U - S$, the fibre $f^{-1}(u)$ is reduced. As $f^{-1}(u)$ is connected and reduced for $u \in U'$, g has the same value on $f^{-1}(u)$. Hence there exists a continuous function h' on U' such that $g = h' \circ f$ on $f^{-1}(U')$. Let $\pi : \tilde{X} \longrightarrow X$ be a desingularization of X. We set $\tilde{f} = f \circ \pi$. There exists a

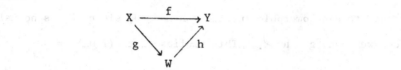

nowhere dense analytic subset T of U' such that for any point $y \in U'' = U' - T$ there exist a neighbourhood U_y of y in U'' and a holomorphic section $o_y : U_y \longrightarrow \tilde{f}^{-1}(U_y)$ (see Corollary 1.7). Then $h'|_{U_y}$ is equal to $g \circ \pi|_{o_y(U_y)}$ via the isomorphism o_y. Hence h' is holomorphic on U'' and by virtue of normality of U', h' is holomorphic on U' (see Grauert and Remmert [1], Satz 22, p.286). Since g is bounded on $f^{-1}(U)$, h' is bounded on U'. Hence h' can be extended to a holomorphic function h on U, since U is normal. Hence, we have $g = h \circ f$. This implies that $(f_* \underline{O}_X)_y = \underline{O}_{Y,y}$.

Q.E.D.

Corollary 1.14. Let $f : X \longrightarrow Y$ be a proper modification (see Definition 2.1, below) of normal complex spaces. Then

$$f_* \underline{O}_X = \underline{O}_Y .$$

Proof. By virtue of Corollary 1.12, every fibre of the morphism f is connected. Hence by Proposition 1.13 we have the desired result.

Q.E.D.

§ 2. Meromorphic mappings and the resolutions of singularities

All analytic spaces that we shall consider in this section are assumed to be irreducible and reduced (but not necessarily compact) and a complex (algebraic) varieties are assumed to be compact, unless otherwise explicitly mentioned.

From among the various definitions of meromorphic mappings, we choose the one due to Remmert [2]. The reader is referred to Remmert [2] and Stein [1], for a more general discussion. First we shall introduce the notion of a proper modification.

Definition 2.1. A morphism $f : X \longrightarrow Y$ of complex spaces is called a proper modification, if

1) f is proper and surjective ;

2) there exist nowhere dense analytic subsets M of X and N of Y such that f induces a biholomorphic mapping of $X - M$ onto $Y - N$.

If X and Y are compact, that is, X and Y are complex varieties, a proper modification $f : X \longrightarrow Y$ is often called simply a modification.

Definition 2.2. Let X and Y be complex spaces. A mapping φ of X into the power set of Y is called a meromorphic mapping of X into Y (we shall write $\varphi : X \longrightarrow Y$), if X satisfies the following conditions.

1) The graph $G_\varphi = \{(x, y) \in X \times Y \mid y \in \varphi(x)\}$ of φ is an irreducible analytic subset in $X \times Y$.

2) The projection map $p_X : G_\varphi \longrightarrow X$ is a proper modification.

Let $p_Y : G_\varphi \longrightarrow Y$ be the projection map to the second factor. For a subset A of X we define $\varphi(A) = p_Y(p_X^{-1}(A))$. For a subset B of Y we define $\varphi^{-1}(B) = p_X(p_Y^{-1}(B))$.

For a meromorphic mapping $\varphi : X \longrightarrow Y$, there exists a smallest nowhere dense analytic subset $S(\varphi)$ of X such that φ induces a morphism of $X - S$ into Y. This analytic subset $S(\varphi)$ is called the <u>set of points of indeterminacy</u> of φ. Then $p_X : G_\varphi \longrightarrow X$ induces an isomorphism between $G_\varphi - p_X^{-1}(S(\varphi))$ and $X - S(\varphi)$.

Conversely, suppose that there exists an irreducible analytic subset G in $X \times Y$ which satisfies the condition :

(M) The projection $p_X : G \longrightarrow X$ is a proper modification.

Now, if $p_Y : G \longrightarrow Y$ is the projection to the second factor and we set $\varphi(x) = p_Y \circ p_X^{-1}(x)$, then φ enjoys properties 1) and 2) of Definition 2.2. Hence φ is a meromorphic mapping and one can check easily that the graph G_φ of φ in $X \times Y$ is G.

In summary, a meromorphic mapping is uniquely determined by an analytic subset of $X \times Y$ which satisfies the condition (M). Note that a meromorphic mapping $\varphi : X \longrightarrow Y$ is a morphism if and only if $p_X : G_\varphi \longrightarrow X$ is an isomorphism.

<u>Remark 2.3</u>. When X and Y are algebraic, to give an irreducible algebraic subset G in $X \times Y$ satisfying condition (M) above is equivalent to provide a rational map. The key references in this connection are Weil [1], Lang [1] and Zariski [1], [4]. Therefore, by GAGA (Theorem 1.3), if X and Y are complete algebraic varieties, the meromorhic mappings from X to Y are precisely the rational mappings from X to Y.

The following examples of meromorphic mappings play a very

important role in our theory.

Example 2.4.1. Let $f_1, f_2, \cdots, f_N$ be meromorphic functions on a complex variety X. Suppose that S is a nowhere dense analytic subset of X such that, on $X - S$, $f_1, f_2, \cdots, f_N$ are holomorphic. We set $X' = X - S$ and $G' = \{(x, f_1(x), f_2(x), \ldots, f_N(x)) \in X' \times \mathbb{C}^N \mid x \in X'\}$. G' is the graph of the holomorphic mapping

$$\Psi' : \begin{array}{ccc} X' & \longrightarrow & \mathbb{C}^N \\ \cup & & \cup \\ x & \longmapsto & (f_1(x), f_2(x), \ldots, f_N(x)). \end{array}$$

$X' \times \mathbb{C}^N$ can be imbedded in a natural way into $X \times \mathbb{P}^N$. Let G be the closure of G' in $X \times \mathbb{P}^N$. We can prove that G is an analytic subset of $X \times \mathbb{P}^N$ which satisfies the above condition (M). Hence, G defines a meromorphic map Φ, which we shall write

$$\Phi : \begin{array}{ccc} X & \longrightarrow & \mathbb{P}^N \\ \cup & & \cup \\ x & \longmapsto & (1 : f_1(x) : f_2(x) : \cdots : f_N(x)). \end{array}$$

For more details, the reader is advised to see Remmert [2] and Kuhlmann [1].

Example 2.4.2. Suppose that $\varphi_0, \varphi_1, \varphi_2, \cdots \varphi_N$ are linearly independent global sections of a complex line bundle L on a complex variety X. We set

$$S = \{x \in X \mid \varphi_0(x) = 0, \ \varphi_1(x) = 0, \ \varphi_2(x) = 0, \cdots, \ \varphi_N(x) = 0\}.$$

S is a nowhere dense analytic subset of X. Let G' be the graph of the holomorphic mapping

$$\Phi'_L : \begin{array}{ccc} X' = X - S & \longrightarrow & \mathbb{P}^N \\ \cup & & \cup \\ x & \longmapsto & (\varphi_0(x) : \varphi_1(x) : \varphi_2(x) : \cdots : \varphi_N(x)), \end{array}$$

and let G be the closure of G' in $X \times \mathbb{P}^N$. Since

$$\frac{\varphi_1}{\varphi_0}, \frac{\varphi_2}{\varphi_0}, \ldots, \frac{\varphi_N}{\varphi_0}$$

are meromorphic, we can use the results of Example 2.4.1 and conclude that G is an analytic subset of $X \times \mathbb{P}^N$ which satisfies the condition (M). Hence an analytic set G provides us with a meromorphic mapping Φ_L which we shall write

$$\begin{array}{ccc} \Phi_L : X & \longrightarrow & \mathbb{P}^N \\ \cup\!\!\!\shortmid & & \cup\!\!\!\shortmid \\ x & \longmapsto & (\varphi_0(x) : \varphi_1(x) : \varphi_2(x) : \cdots : \varphi_N(x)). \end{array}$$

This map will be more closely examined in §5 and §7.

Remmert [2] has proved the following valuable result.

<u>Theorem 2.5.</u> If X and Y are normal complex varieties and $\varphi : X \longrightarrow Y$ is a meromorphic mapping, the set $S(\varphi)$ of points of indeterminacy of the meromorphic mapping φ is an analytic subset of X of codimension at least two.

<u>Definition 2.6.</u> A meromorphic mapping $\varphi : X \longrightarrow Y$ is called <u>generically surjective</u> if the projection

$$p_Y : G_\varphi \longrightarrow Y$$

induced by the projection of $X \times Y$ onto the second factor is surjective.

Let $\varphi : X \longrightarrow Y$ and $\psi : Y \longrightarrow Z$ be meromorphic mappings. The composition $\psi \circ \varphi$ does not necessarily exist. However, if φ is generically surjective, then we can define the composition in the following way. For a point $x \in X' = X - (S(\varphi) \cup \varphi^{-1}(S(\psi)))$, we set $\psi \circ \varphi(x) = \psi(\varphi(x))$. For a point $x \in X - X'$, $\psi \circ \varphi(x)$ consists of all points $z \in Z$ such that there exist a point $y \in Y$ and a sequence $\{x_\nu\}$, $\nu = 1, 2, \ldots,$ in X' such that $\{x_\nu\}$, $\{\varphi(x_\nu)\}$, $\{\psi(\varphi(x_\nu))\}$

convergent to x, y and z in X, Y and Z, respectively. Then $\psi \circ \varphi$ defines a meromorphic mapping of X into Z. For a point $x \in X - X'$, $\psi \circ \varphi(x)$ may not be equal to $\psi(\varphi(x))$.

Definition 2.7. A meromorphic mapping $\varphi : X \longrightarrow Y$ of complex varieties is called a bimeromorphic mapping if $p_Y : G \longrightarrow Y$ is also a proper modification.

If φ is a bimeromorphic mapping, the analytic set

$$\{(y, x) \in Y \times X \mid (x, y) \in G_\varphi\} \subset Y \times X$$

defines a meromorphic mapping $\varphi^{-1} : Y \longrightarrow X$ such that $\varphi \circ \varphi^{-1} = id_Y$ and $\varphi^{-1} \circ \varphi = id_X$.

Two complex varieties X and Y are called bimeromorphically equivalent if there exists a bimeromorphic mapping $\varphi : X \longrightarrow Y$.

In order to study a meromorphic mapping Φ_L in Example 2.4.2 more closely, we need the concept of a projective fibre space $\mathbb{P}(\underline{F})$ associated with a coherent sheaf $\underline{F}$ on an analytic space X.

First we shall give a functor theoretic definition of a projective fibre space. The reader is referred to Sém. Cartan 60/61, Exp. 12, for a more general discussion.

(2.8) Let (An)/X be the category of analytic spaces over a complex space X. (Here analytic spaces may not be irreducible nor reduced.) Let $\underline{F}$ be a coherent sheaf on X. For any element $f : Y \longrightarrow X$ of (An)/X, $\underline{\mathbb{P}}(\underline{F})(Y)$ is defined as the set of invertible sheaves on Y which are quotient of $f^*\underline{F}$. Then $\underline{\mathbb{P}}(\underline{F})$ is a contravariant functor of (An)/X into (Set). The functor $\underline{\mathbb{P}}(\underline{F})$ can be represented by an element $p : \mathbb{P}(\underline{F}) \longrightarrow X$ of (An)/X. We call $p : \mathbb{P}(\underline{F}) \longrightarrow X$ a projective fibre space associated with $\underline{F}$. Notice that $\mathbb{P}(\underline{F})$ may not be reduced, even if X is reduced.

If $\underline{F} = \underline{O}_X^{n+1}$, the free sheaf of rank $n+1$, then the associated projective fibre space is nothing other than $p : X \times \mathbb{P}^n \longrightarrow X$ where p is the projection map. More generally suppose that $\underline{F}$ is a locally free sheaf of rank $n+1$ on X, which corresponds to a vector bundle F on X. Then the projective fibre space $p : \mathbb{P}(\underline{F}) \longrightarrow X$ is analytically isomorphic to the projective bundle associated with the dual bundle F^* of F. (If $\{f_{ij}\}$, $f_{ij}(z) \in GL(n+1, \mathbb{C})$, is a set of transition functions of the vector bundle F^*, we consider $f_{ij}(z)$ as elements of $PGL(n, \mathbb{C})$ and construct a projective bundle on X. This projective bundle is called the projective bundle associated with F^*.)

Let $f : Y \longrightarrow X$ be a morphism of analytic spaces. Then, by the definition of the functor $\mathbb{P}(\underline{F})$, we can easily show that there is a canonical isomorphism

$$\mathbb{P}(f^*\underline{F}) \xrightarrow{\sim} \mathbb{P}(\underline{F}) \times_X Y .$$

Hence, for a projective fibre space $p : \mathbb{P}(\underline{F}) \longrightarrow X$ and for any point $x \in X$, the fibre $p^{-1}(x) = \mathbb{P}(\underline{F})_x$ is canonically isomorphic to $\mathbb{P}(\underline{F}_{|x})$, where $\underline{F}_{|x} = \underline{F}_x \otimes_{\underline{O}_{X,x}} \mathbb{C}$. Hence the fibre $\mathbb{P}(\underline{F})_x$ is empty or a complex projective space.

Next we shall give an elementary but non-intrinsic construction of $p : \mathbb{P}(\underline{F}) \longrightarrow X$.

(2.9). Let $\{U_i\}_{i \in I}$ be a small Stein open covering of X such that on each U_i there exists an exact sequence

(2.9.1). $$\underline{O}^p_{X|U_i} \xrightarrow{\alpha} \underline{O}^q_{X|U_i} \longrightarrow \underline{F}_{|U_i} \longrightarrow 0$$

of sheaves. The $\underline{O}_X$-linear homomorphism α can be represented by a

$q \times p$ matrix $M(z) = (m_{ij}(z))$ such that

$$\begin{pmatrix} a_1(z) \\ \cdot \\ \cdot \\ \cdot \\ a_q(z) \end{pmatrix} = M(z) \cdot \begin{pmatrix} b_1(z) \\ \cdot \\ \cdot \\ \cdot \\ b_p(z) \end{pmatrix} .$$

Consider the product $U_i \times \mathbb{P}^{q-1}$ where $(\xi_1 : \cdots : \xi_q)$ are homogeneous coordinates of $\mathbb{P}^{q-1}$. We define $\mathbb{P}(\underline{F})_{|U_i}$ to be the analytic subspace (possibly empty) of $U_i \times \mathbb{P}^{q-1}$ defined by the equations

(2.9.2). $$\sum_{i=1}^{q} m_{ij}(z) \cdot \xi_i = 0, \quad j = 1, 2, \cdots, p.$$

The construction of the $\mathbb{P}(\underline{F}_{|U_i})$ depends on our choice of a resolution (2.9.1). One can prove however that the structure of the analytic space $\mathbb{P}(\underline{F}_{|U_i})$ depends only on the coherent sheaf $\underline{F}_{|U_i}$, by using a chain homotopy property of free resolution of a module (see Cartan and Eilenberg [1], p.75-78). Hence, we can patch together the $\mathbb{P}(\underline{F}_{|U_i})$, $i \in I$, and obtain in this way the analytic space $\mathbb{P}(\underline{F})$.

The projection $pr_2 : U_i \times \mathbb{P}^{q-1} \longrightarrow U_i$ induces the structure morphism $p : \mathbb{P}(\underline{F}) \longrightarrow X$. Suppose that X is an algebraic variety and $\underline{F}$ is an algebraic coherent sheaf on X. Then similarly we can define the algebraic projective fibre space $p : \mathbb{P}(\underline{F}) \longrightarrow X$ associated with the algebraic coherent sheaf $\underline{F}$ (see EGA I_a, 9.7). Then, by GAGA, we can easily prove that $p^{an} : \mathbb{P}(\underline{F})^{an} \longrightarrow X^{an}$ is canonically isomorphic to the projective fibre space $\mathbb{P}(\underline{F}^{an})$ over X^{an}. Later we shall often use this fact.

In the later parts of this book we shall often encounter the situation described below.

(2.10). Let $f : X \longrightarrow Y$ be a proper surjective morphism of complex varieties and let L be a line bundle on X such that $\dim_{\mathbb{C}} H^0(X, \underline{O}(L)) \geq 2$. We set $\underline{F} = f_* \underline{O}(L)$. By Grauert's proper mapping theorem (1.4), again, $\underline{F}$ is a coherent sheaf and we can construct the projective fibre space $p : \mathbb{P}(\underline{F}) \longrightarrow X$ associated with $\underline{F}$.

If U is a small Stein open set of Y, then $\underline{F}_{|U}$ is spanned by $\varphi_1, \varphi_2, \cdots, \varphi_q \in H^0(U, \underline{F})$. Through the canonical isomorphism

$$H^0(U, \underline{F}) \xrightarrow{\sim} H^0(f^{-1}(U), \underline{O}(L)),$$

we consider $\varphi_1, \varphi_2, \cdots, \varphi_q$ as elements of $H^0(f^{-1}(U), \underline{O}(L))$. Therefore, there is a meromorphic mapping $\hat{\psi}_U$

$$\begin{array}{rcl} \hat{\psi}_U : f^{-1}(U) & \longrightarrow & U \times \mathbb{P}^{q-1} \\ \cup\!\!\! \shortmid & & \cup\!\!\! \shortmid \\ z & \longmapsto & (f(z), (\varphi_1(z) : \varphi_2(z) : \cdots : \varphi_q(z))). \end{array}$$

$\underline{F}_{|U}$ has a resolution (2.9.1). As the $\varphi_1, \varphi_2, \cdots, \varphi_q$ lie in $H^0(U, \underline{F})$, any point of the image of $\hat{\psi}_U$ satisfies the equations (2.9.2). It follows that there is a meromorphic mapping

$$\psi_U : f^{-1}(U) \longrightarrow \mathbb{P}(\underline{F}_{|U})_{red},$$

where $P(\underline{F}_{|U})_{red}$ means the reduced complex space associated with $\mathbb{P}(\underline{F}_{|U})$ Glueing the ψ_U we obtain a <u>meromorphic mapping</u>

$$\psi : X \longrightarrow \mathbb{P}(\underline{F})_{red}$$

which makes the diagram

$$\begin{array}{ccc} X & \xrightarrow{\ \psi\ } & \mathbb{P}(\underline{F})_{red} \\ & {\scriptstyle f}\searrow \quad \swarrow{\scriptstyle p} & \\ & Y & \end{array}$$

commute. For simplicity we often omit "red" and say that there exists a meromorphic mapping $\psi : X \longrightarrow \mathbb{P}(\underline{F})$.

Note that there exists an open dense subset Y' of Y such that

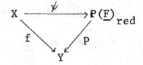

$\underline{O}(L)$ is flat over Y', $\underline{F}_{|Y'}$ is a locally free sheaf of rank $r \geq 2$, since $\dim_{\mathbb{C}} H^0(X, \underline{O}(L)) \geq 2$, and that by Theorem 1.4, 3) we find that, for each point $y \in Y'$, there are a neighbourhood U of y in Y' and elements $\psi_1, \psi_2, \cdots, \psi_r$ of $H^0(f^{-1}(U), \underline{O}(L))$ having the property that, for each point $z \in U$, $\psi_1, \psi_2, \cdots, \psi_r$ induce a basis of $H^0(X_z, \underline{O}(L_z))$, where $X_z = f^{-1}(z)$ and L_z is the restriction of L to X_z. Hence, for any point $y \in Y'$, <u>the restriction</u>

$$\psi_y : X_y \longrightarrow \mathbb{P}(\underline{F})_y$$

<u>of the meromorphic mapping</u> ψ <u>on</u> X_y <u>is nothing other than the meromorphic mapping</u> Φ_{L_y} in Example 2.4.2. The last remark is frequently employed later.

Resolution of singularities in the complex case in its most general form is due to Hironaka. We present some of the key results in this direction. First we recall the definition of a monoidal transformation.

<u>Definition 2.11</u>. Let X be an analytic space, D an analytic subspace (not necessarily reduced) and $\underline{J}$ the ideal sheaf of D on X. A pair (D, f) consisting of the analytic subspace D and a morphism of analytic spaces $f : X' \longrightarrow X$ is called the <u>monoidal transformation of</u> X <u>with center</u> D when the morphism f satisfies the following conditions.

1) The ideal sheaf generated by the image of $f^{-1}(\underline{J}) \otimes \underline{O}_{X'} \longrightarrow \underline{O}_{X'}$ in $\underline{O}_{X'}$ is invertible on X'.

2) If $g : X'' \longrightarrow X$ is a morphism of analytic spaces having the property 1), there exists a unique morphism $h : X'' \longrightarrow X'$ of analytic spaces so that $g = f \circ h$.

The existence of monoidal transformations is guaranteed by the results to be found in Hironaka [1] and Hironaka and Rossi [1]. Here, one finds that if the analytic space X is reduced, so is the analytic space X' obtained by a monoidal transform (D, f). Note that a monoidal transformation is a proper modification.

<u>Theorem 2.12</u>. Let X be a complex variety. Then there exists a sequence of monoidal transformations

$$f_i : X_i \longrightarrow X_{i-1}, \quad i = 1, 2, \cdots , n ,$$

which satisfies the following conditions :

1) $X_0 = X$ and X_n is non-singular ;

2) the center of the monoidal transformation f_i is non-singular and contained in the singular locus of X_{i-1}.

Moreover if a finite group G operates on X, then we can choose the above sequence of monoidal transformations in such a way that the group G can be lifted to a group of analytic automorphisms of X_n.

The proof and the more precise formulation of this theorem is found in Hironaka [1], [2] and Lejeune and Tessier [1].

<u>Theorem 2.13</u>. (Elimination of points of indeterminacy of a rational mapping) Consider a rational mapping $g : X \longrightarrow Y$ of algebraic varieties. By finite succession of monoidal transformations with non-singular centers, we have an algebraic variety $\widetilde{X}$ and a proper modification $h : \widetilde{X} \longrightarrow X$ such that $g \circ h : \widetilde{X} \longrightarrow Y$ is a morphism.

For the proof, see Hironaka [1].

The following proposition is a corollary of local Chow's lemma and Main Theorem II' in Hironaka [1] (see Hironaka [1], p.152-153). Local Chow's lemma is a corollary of the flatterning theorem due to

Hironaka [4].

Proposition 2.14. Let $f : X \longrightarrow Y$ be a proper modification of complex manifolds. Then we have

$$R^p f_* \underline{O}_X = 0$$

for any positive integer p.

Corollary 2.15. If two complex manifolds X and Y are bimeromorphically equivalent, then

$$\dim H^p(X, \underline{O}_X) = \dim H^p(Y, \underline{O}_Y) .$$

Proof. Let $\varphi : X \longrightarrow Y$ be a meromorphic mapping and let G_φ be its graph. If $\tilde{X}$ is a non-singular model of G_φ, then the canonical morphisms $f : \tilde{X} \longrightarrow X$ and $g : \tilde{X} \longrightarrow Y$ are modifications. Hence it is enough to consider the case where φ is a modification. Then we have a spectral sequence

$$E_2^{p,q} = H^p(Y, R^q \varphi_* \underline{O}_X) \Longrightarrow H^{p+q}(X, \underline{O}_X) .$$

By Proposition 2.14 the spectral sequence degenerates. Hence we have the desired result. Q.E.D.

§3. Algebraic dimensions and algebraic reductions of complex varieties

Let $\mathbb{C}(M)$ be the field of meromorphic functions on a complex variety M. If two complex varieties M_1 and M_2 are bimeromorphically equivalent, then $\mathbb{C}(M_1)$ and $\mathbb{C}(M_2)$ are isomorphic. Hence, if the meromorphic function field $\mathbb{C}(M)$ of a complex variety M is always finitely generated over $\mathbb{C}$, then the transcendence degree of $\mathbb{C}(M)$ defines a bimeromorphic invariant. The following theorem assures that this is the case.

Theorem 3.1. If M is a complex variety, the field $\mathbb{C}(M)$ of meromorphic functions on M is a finitely generated extension over $\mathbb{C}$(an algebraic function field over $\mathbb{C}$) satisfying

$$\text{tr. deg.}_{\mathbb{C}}\mathbb{C}(M) \leqq \dim M .$$

Therefore, there exists a projective variety V such that $\dim V \leqq \dim M$ and $\mathbb{C}(M)$ is isomorphic to the rational function field $\mathbb{C}(V)$ of V.

The proof is found in Thimm [1], Remmert [1] and Andreotti and Stoll [1].

Definition 3.2. The algebraic dimension $a(M)$ of a complex variety M is defined by

$$a(M) = \text{tr.deg.}_{\mathbb{C}}\mathbb{C}(M) .$$

Theorem 3.1 implies that $a(M) \leqq \dim M$. Moreover, $a(M) = 0$ if and only if $\mathbb{C}(M) = \mathbb{C}$, i.e., M has no non-constant meromorphic functions. By a resolution of singularities (Theorem 2.22), the projective variety V which appears in Theorem 3.1 can be chosen to be smooth. Let $\mathbb{C}[V] = \mathbb{C}[\zeta_0, \zeta_1, \ldots, \zeta_N]$ be a homogeneous coordinate ring of V.

Then

$$\frac{\zeta_1}{\zeta_0}, \frac{\zeta_2}{\zeta_0}, \dots, \frac{\zeta_N}{\zeta_0}$$

are rational functions on V. Hence, under the isomorphism between $\mathbb{C}(M)$ and $\mathbb{C}(V)$ of Theorem 3.1, there are meromorphic functions $\varphi_1, \varphi_2, \dots, \varphi_N$ corresponding to $\frac{\zeta_1}{\zeta_0}, \frac{\zeta_2}{\zeta_0}, \dots, \frac{\zeta_N}{\zeta_0}$, respectively.

Suppose that $\Psi : M \longrightarrow \mathbb{P}^N$ is the meromorphic mapping defined by

$$M \ni z \longmapsto (1 : \varphi_1(z) : \varphi_2(z) : \cdots : \varphi_N(z)) \in \mathbb{P}^N$$

(see 2.4.1). Let G be the graph of this meromorphic mapping. Consider the surjective morphism $\pi : M^* \longrightarrow G$ obtained by a resolution of the singularities of G. The morphism $\Psi^* : M^* \longrightarrow V$ obtained from the composition of π with the projection p_V of G onto V,

$$\Psi^* : M^* \xrightarrow{\ \pi\ } G \xrightarrow{\ p_V\ } V ,$$

has the following properties.

1) M^* is smooth and bimeromorphically equivalent to M.

2) V is a projective manifold and $\dim V = a(M)$.

3) Ψ^* induces an isomorphism between $\mathbb{C}(V)$ and $\mathbb{C}(M)$.

Definition 3.3. A surjective morphism

$$\Psi^* : M^* \longrightarrow V$$

is called an algebraic reduction of a complex variety M if it satisfies the above conditions 1), 2) and 3).

An algebraic reduction of a complex variety M is unique up to a bimeromorphic equivalence. This means that if $\Psi_1^* : M_1^* \longrightarrow V_1^*$ is another algebraic reduction, there exist bimeromorphic mappings $f : V \longrightarrow V_1$ and $f^* : M^* \longrightarrow M_1^*$ such that $\Psi_1^* \circ f^* = f \circ \Psi^*$.

An *analytic fibre space* or, simply, a *fibre space* is a morphism $g : X \longrightarrow Y$ of reduced analytic spaces, where

1) g is proper and surjective;

2) the general fibre of g is connected.

The general fibre of an algebraic reduction $\Psi^* : M^* \longrightarrow V$ is smooth, by Corollary 1.8. The following proposition implies that an algebraic reduction is a fibre space.

Proposition 3.4. The fibres of an algebraic reduction $\Psi^* : M^* \longrightarrow V$ are connected. Hence, if $a(M) = \dim M$, Ψ^* is a modification.

The proof is an easy consequence of Corollary 1.10.

The structure of the fibres of an algebraic reduction $\Psi^* : N^* \longrightarrow V$ will be studied in §12.

Definition 3.5. A complex variety (manifold) is called a *Moishezon variety (manifold)* if $a(M) = \dim M$.

The following theorem due to Moishezon shows that a Moishezon manifold is not so far from a projective manifold.

Theorem 3.6. Let M be a Moishezon variety. Then there exists a smooth projective variety M^* and a modification $f : M^* \longrightarrow M$ such that f is obtained by finite succession of monoidal transformations with non-singular centers.

A proof of this theorem is found in Moishezon [1], Chapter II.

Remark 3.7. 1) Artin has shown that any Moishezon variety carries an algebraic structure, that is, any Moishezon variety is a proper algebraic space over $\mathbb{C}$ (see Artin [1]).

2) A Moishezon manifold M is projective algebraic if and only

if M carries a Kähler matric (see Moishezon [1], Chapter I, Theorem 11.).

Theorem 3.8. 1) Let $f : M \longrightarrow N$ be a fibre space of complex varieties. Then there exists a nowhere dense analytic subset N_1 of N such that

$$a(N) \leqq a(M) \leqq a(N) + a(M_y) \leqq a(N) + \dim f ,$$

for any $y \in N_1$, where $M_y = f^{-1}(y)$ is the fibre of f over y and $\dim f = \dim M - \dim N$.

2) When M is a subvariety of a complex variety N, we have

$$a(N) \leqq a(M) + \operatorname{codim} M.$$

Theorem 3.8 will be proved in § 7.

Corollary 3.9. 1) Subvarieties of a Moishezon variety are Moishezon varieties.

2) If M is a Moishezon variety and $f : M \longrightarrow N$ is a surjective morphism of complex varieties, then N is a Moishezon variety.

Corollary 3.10. Let $f : M \longrightarrow N$ be a fibre space of complex varieties. Suppose that $a(N) = \dim N$ and that some fibre $M_y = f^{-1}(y)$, $y \in N$ has an irreducible component D satisfying $a(D) = 0$. Then $a(M) = a(N)$.

Proof. Assume that $a(M) > a(N)$. Clearly, $\dim D \geqq \dim f$. Applying Theorem 3.8,2) to D, we have

$$\begin{aligned} a(D) &\geqq a(M) - \operatorname{codim} D = a(M) - (\dim M - \dim D) \\ &\geqq a(M) - (\dim M - \dim f) \\ &\geqq a(M) - \dim N > a(N) - \dim N = 0. \end{aligned}$$

Hence, $a(D) > 0$. This is a contradiction. Q.E.D.

Chapter II

D-dimensions and Kodaira dimensions

The main purpose of the present chapter is to introduce the notion of D-dimensions and Kodaira dimensions.

For that purpose, in §4, some important results on divisors and linear systems on a complex variety are given. First we shall give the definitions of Cartier divisors and Weil divisors and study their relationship. Next a linear system associated with a Cartier divisor is studied and theorems of Bertini are proved. The results in this section are classical and well-known. For our classification theory, we shall need a geometric analysis of linear systems. Therefore we use the classical geometric approach to the theory of linear systems.

In §5, for a Cartier divisor D on a complex variety V, the D-dimension $\kappa(D, V)$, of the divisor D is defined. This is, roughly speaking, the maximal dimension of the image varieties of meromorphic mappings Φ_{mD} corresponding to the divisor mD (see Definition 5.1, below). The notion of D-dimension is due to Iitaka (see Iitaka [2]). For a line bundle L on a complex variety V, the corresponding notion of L-dimension $\kappa(L, V)$ is easily introduced. In this section the elementary properties of D-dimensions (L-dimensions) are given. The fundamental theorem on D-dimensions (Theorem 5.10, below) due to Iitaka is also stated here. The proof will be given in Chapter III, §7. Almost all results in this section are due to Iitaka.

In §6, the Kodaira dimension $\kappa(V)$ of a complex variety V is defined. $\kappa(V)$ is, by definition, the L-dimension of the canonical line bundle of a non-singular model of V. The Kodaira dimension is a bimeromorphic invariant. Some fundamental properties of Kodaira dimensions are given. Some of them are immediate consequences of the corresponding results on D-dimensions. The fundamental theorem on the pluricanonical fibrations (Theorem 6.11) is also an almost immediate consequence of Theorem 5.10. The results in this section are also mostly due to Iitaka. The application of the fundamental theorem to classification theory will be discussed in Chapter IV.

§ 4. Divisors and linear systems

Definition 4.1. Let V be a complex variety. A Cartier divisor on V is a global section of the sheaf

$$\underline{M}_V^* / \underline{O}_V^*$$

where $\underline{M}_V^*$ is the sheaf of germs of not identically vanishing meromorphic functions on V, and $\underline{O}_V^*$ is the sheaf of germs of nowhere vanishing holomorphic functions on V.

If $\{U_i\}_{i\in I}$ is an appropriate open covering of V, a Cartier divisor D on V is a collection of meromorphic functions φ_i on U_i, $i \in I$, such that on $U_i \cap U_j \neq \emptyset$, $\frac{\varphi_i}{\varphi_j}$ and $\frac{\varphi_j}{\varphi_i}$ are holomorphic. We call the meromorphic function φ_i a local equation of D on U_i. Note that the open covering $\{U_i\}_{i\in I}$ depends on D. A local equation of D on U_i is unique up to elements of $H^0(U_i, \underline{O}_V^*)$.

The multiplicative group structure of the sheaf $\underline{M}_V^*/\underline{O}_V^*$ induces an additive group structure on the set of all Cartier divisors. Let D_1 and D_2 be Cartier divisors with local equations $\{\varphi_i\}_{i\in I}$, $\{\psi_i\}_{i\in I}$ on $\{U_i\}_{i\in I}$, respectively. Then the local equations $\{\varphi_i \cdot (\psi_i)^{\pm 1}\}_{i\in I}$ on $\{U_i\}_{i\in I}$ determine a Cartier divisor $D_1 \pm D_2$. The zero element of this module (we call it the zero (or trivial) Cartier divisor) is the one whose local equation is 1 on any open set.

With any Cartier divisor D we can associate a coherent sheaf $\underline{O}_V(D)$ defined by

$$\underline{O}_V(D)_x = \varphi_i^{-1} \cdot \underline{O}_{V,x} \subset \underline{M}_V, \quad x \in U_i ,$$

where φ_i is a local equation of D on U_i and $\underline{M}_V$ is the sheaf of germs of meromorphic functions on V. It is clear that $\underline{O}_V(D)$ is locally isomorphic to $\underline{O}_V$, and, hence, is an invertible sheaf. Note that

$D = 0$ if and only if $\underline{O}_V(D) = \underline{O}_V$ as $\underline{O}_V$-ideals.

Definition 4.2. A non-zero Cartier divisor D on V is called effective (we write $D > 0$) if all the local equations of D are holomorphic functions and at least one local equation has zeros. For two Cartier divisors D_1, D_2, if $D_1 - D_2 > 0$, we write $D_1 > D_2$. We write $D_1 \geqq D_2$, if $D_1 > D_2$ or $D_1 = D_2$.

The proof of the following lemma is left to the reader.

Lemma 4.3. The three conditions described below are equivalent .

1) $D > 0$.

2) $\underline{O}_V \subsetneqq \underline{O}_V(D)$.

3) $\underline{O}_V(-D)$ is a sheaf of $\underline{O}_V$ - ideals such that $\underline{O}_V(-D) \neq \underline{O}_V$.

Definition 4.4. The support of a Cartier divisor D on V is the subset (analytic or algebraic) of all points $x \in V$ such that 1 can not be taken as a local equation of D at x.

A local equation φ_i of a Cartier divisor D on the open set U_i can be viewed as a quotient

$$\varphi_i = f_i / g_i$$

where f_i and g_i are holomorphic on U_i. The support of D in U_i consists of some of the components of

$$f_i \cdot g_i = 0$$

on U_i. The local rings of V are Noetherian (see for example, Cunning and Rossi [1], p.72). Hence, applying the principal ideal theorem (see for example, Zariski and Sammuel [1], Theorem 29, p.238), we see that the components of $f_k \cdot g_i = 0$ on U_i are of pure codimension one in U_i. It follows that the support of D on V is an analytic subset of V of pure codimension one, if $D \neq 0$.

Consider the short exact sequence of sheaves

$$0 \longrightarrow \underline{O}_V^* \longrightarrow \underline{M}_V^* \longrightarrow \underline{M}_V^*/\underline{O}_V^* \longrightarrow 0$$

on V. From this exact sequence, we obtain a long exact sequence of abelian groups

$$(4.5) \quad 0 \longrightarrow H^0(V, \underline{O}_V^*) \longrightarrow H^0(V, \underline{M}_V^*) \xrightarrow{\alpha} H^0(V, \underline{M}_V^*/\underline{O}_V^*) \longrightarrow$$

$$\xrightarrow{\beta} H^1(V, \underline{O}_V^*) \longrightarrow H^1(V, \underline{M}_V^*) \longrightarrow \cdots .$$

This long exact sequence will be an important tool in our further study of Cartier divisors.

(4.6) We discuss here Cartier divisors in so far as they are defined either in $(Sch/\mathbb{C})$ or (An).

Suppose that V is an algebraic variety.

There are two different types of sheaves on V, accordingly as V has an algebraic or analytic structure :

$$(\underline{O}_V^{alg})^*, \ (\underline{M}_V^{alg})^* \quad \text{in the Zariski topology,}$$

$$(\underline{O}_V^{an})^*, \ (\underline{M}_V^{an})^* \quad \text{in the complex topology.}$$

The functor ψ in §1, induces a mapping of long exact sequences

$$\begin{array}{ccccccc} 0 \longrightarrow H^0(V, (\underline{O}_V^{alg})^*) & \longrightarrow & H^0(V, (\underline{M}_V^{alg})^*) & \longrightarrow & H^0(V, (\underline{M}_V^{alg})^*/(\underline{O}_V^{alg})^*) \\ \downarrow & & \downarrow & & \downarrow \\ 0 \longrightarrow H^0(V, (\underline{O}_V^{an})^*) & \longrightarrow & H^0(V, (\underline{M}_V^{an})^*) & \longrightarrow & H^0(V, (\underline{M}_V^{an})^*/(\underline{O}_V^{alg})^*) \end{array}$$

$$\begin{array}{ccccc} \longrightarrow H^1(V, (\underline{O}_V^{alg})^*) & \longrightarrow & H^1(V, (\underline{M}_V^{alg})^*) & \longrightarrow & \cdots\cdots\cdots \\ \downarrow & & \downarrow & & \\ \longrightarrow H^1(V, (\underline{O}_V^{an})^*) & \longrightarrow & H^1(V, (\underline{M}_V^{an})^*) & \longrightarrow & \cdots\cdots\cdots \end{array}$$

As V is complete, we have

$$H^0(V, (\underline{O}_V^{alg})^*) = H^0(V, (\underline{O}_V^{an})^*) = \mathbb{C}^* ,$$

and

$$H^0(V, (\underline{M}_V^{alg})^*) = H^0(V, (\underline{M}_V^{an})^*) = \mathbb{C}(M)^* .$$

Hence the first two arrows which appear in the above mapping of long exact sequences are isomorphisms. Moreover, by GAGA (Theorem 1.3,3), the category of algebraic invertible sheaves on V is equivalent to the category of analytic invertible sheaves on V. This implies that the natural homomorphism

$$H^1(V, (\underline{O}_V^{alg})^*) \longrightarrow H^1(V, (\underline{O}_V^{an})^*)$$

is an isomorphism. ($H^1(V, (\underline{O}_V^{alg})^*)$ and $H^1(V, (\underline{O}_V^{an})^*)$ are just the isomorphism classes of algebraic invertible sheaves and analytic invertible sheaves on V, respectively.) Therefore, the fourth arrow in the long exact sequence is an isomorphism. On the other hand, the sheaf $(\underline{M}^{alg})^*$ is flabby, since we have

$$(\underline{M}_V^{alg})^*(U) = \mathbb{C}(V)^*,$$

for any open set U of V. (Note that we always assume that algebraic variety is irreducible.) Hence, from the definition of the sheaf cohomology groups, we have

$$H^1(V, (\underline{M}_V^{alg})^*) = 0.$$

Hence the fifth arrow is injective. The five lemma implies the following.

<u>Lemma 4.7</u>. For an algebraic variety V, there is a natural isomorphism between the group of Cartier divisors on V algebraically defined and the group of Cartier divisors on V analytically defined. We sometimes say, in this situation, that the definitions of Cartier divisors in (Sch/$\mathbb{C}$) and (An) coincide relative to an algebraic variety.

From the above proof of Lemma 4.7 we obtain the following:

<u>Corollary 4.8</u>. The homomorphism

$$\beta : H^0(V, \underline{M}_V^* / \underline{O}_V^*) \longrightarrow H^1(V, \underline{O}_V^*)$$

of the long exact sequence (4.5) is surjective if V is an algebraic variety.

Remark 4.9. 1) If V is not algebraic, the homomorphism β is not necessarily surjective. For example, consider a two-dimensional complex torus with a period matrix

$$\begin{pmatrix} 1 & 0 & \sqrt{-2} & \sqrt{-5} \\ 0 & 1 & \sqrt{-3} & \sqrt{-7} \end{pmatrix} .$$

Siegel has shown that $\mathbb{C}(T) = \mathbb{C}$ and T contains no curves (see Siegel [1], p.104-106). Hence the only Cartier divisor on T must be the trivial Cartier divisor. Consider the exact sequence of sheaves

$$0 \longrightarrow \mathbb{Z} \longrightarrow \underline{O}_T \xrightarrow{\exp} \underline{O}_T^* \longrightarrow 0 .$$

The long exact sequence associated with this short exact sequence contains the following terms

$$\mathbb{C} \xrightarrow{\exp} \mathbb{C}^* \longrightarrow H^1(T, \mathbb{Z}) \xrightarrow{\nu} H^1(T, \underline{O}_T) \longrightarrow H^1(T, \underline{O}_T^*) \longrightarrow \cdots\cdots .$$

As exp is a surjection, ν is an injection. Therefore, $H^1(T, \underline{O}_T^*)$ contains $H^1(T, \underline{O}_T)/H^1(T, \mathbb{Z})$. But it is easily seen that this last group can be endowed with the structure of the torus T.

2) Moishezon [1] has demonstrated that when V is a Moishezon variety, the map β is surjective. As every Moishezon variety is an algebraic space (see Remark 3.7, 1)), with some knowlege of algebraic spaces and with arguments similiar to those above, one can also show that β is surjective. However, it is a deep fact that the category of Moishezon variety is equivalent to the category of complete algebraic spaces of finite type over $\mathbb{C}$.

We examine the map $\beta : H^0(V, \underline{M}_V^*/\underline{O}_V^*) \longrightarrow H^1(V, \underline{O}_V^*)$ more closely. Let D be a Cartier divisor on a complex variety V with local equations $\{\varphi_i\}_{i\in I}$ on an appropriate open covering $\{U_i\}_{i\in I}$ of V. Then, $\beta(D)$ is a one-cycle $\{\varphi_{ij}\}_{i,j\in I}$ where $\varphi_{ij} = \frac{\varphi_j}{\varphi_i}$ is holomorphic on $U_i \cap U_j$, $i,j \in I$. The one-cycle $\{\varphi_{ij}\}_{i,j\in I}$ corresponds to a line bundle whose transition functions are $\{\varphi_{ij}\}_{i,j\in I}$ and the invertible sheaf associated with this line bundle is nothing but $\underline{O}_V(-D)$.

For a Cartier divisor D on V, by $[D]$ we shall mean the line bundle on V corresponding to the invertible sheaf $\underline{O}_V(D)$. Let L be a line bundle on V such that

$$\dim H^0(V, \underline{O}(L)) \geqq 1.$$

Let φ be a non-zero element of $H^0(V, \underline{O}(L))$. φ is represented by a one-cocycle $\{\varphi_i\}$ with respect to an open covering $\{U_i\}$ of V. Then the collection $\{\varphi_i\}$ defines a Cartier divisor D. D is effective, if φ_i vanishes at some points in U_i. If this is not the case, D is the trivial Cartier divisor. Moreover the line bundle L is isomorphic to the line bundle $[D]$.

<u>Definition 4.10</u>. Two Cartier divisors D_1 and D_2 are called <u>linearly equivalent</u> (we write $D_1 \sim D_2$) if $\beta(D_1) = \beta(D_2)$, i.e. the invertible sheaves $\underline{O}_V(-D_1)$ and $\underline{O}_V(-D_2)$ or, equivalently, the line bundles $[D_1]$ and $[D_2]$ are isomorphic.

<u>Definition 4.11</u>. The Cartier divisor which is an image of an element $f \in H^0(V, \underline{M}_V^*)$ under the homomorphism α of (4.5) is called a <u>principal divisor</u> and denoted by the symbol (f).

We will now introduce Weil divisors and study their connection with Cartier divisors.

<u>Definition 4.12</u>. A <u>Weil divisor</u> D of a variety V is a finite formal sum

$$\sum_{i=1}^{N} n_i C_i$$

where C_i is an irreducible subvariety of V of codimension one which is not contained in the singular locus of V and n_i is an integer.

The set of all Weil divisors forms a free abelian group. The zero of this module (we call it the zero Weil divisor) is the Weil divisor such that all n_i's are zero. A Weil divisor $D = \sum_{i=1}^{N} n_i C_i$ is called an effective Weil divisor, if $n_i \geqq 0$ and $D \neq 0$. For two Weil divisors D_1 and D_2 we write $D_1 \geqq D_2$, if $D_1 - D_2$ is an effective divisor or the zero divisor.

If V is a normal complex variety, the codimension of the singular locus is shown to be strictly bigger than one (see for example, Abhyankar [1], 45.15, p.434). For a normal variety V, therefore, the assumption that C_i, $i = 1,2, \ldots, N$, are not contained in the singular locus of V, which appears in Definition 4.12, is automatically satisfied.

The support of a Cartier divisor is of pure codimension one. It is natural to suppose that one can order a Weil divisor to every Cartier divisor and to ask when this correspondence is one to one and onto. The following theorem gives an answer to this question.

Theorem 4.13. 1) There is a natural group homomorphism h from the set of all Cartier divisors to the set of all Weil divisors.

2) h is injective if V is normal.

3) h is bijective if V is smooth.

Proof. 1) Let V_{reg} be the open dense subset of V consisting of all the regular, i.e., smooth, poins of V. Suppose that D is a Cartier divisor on V with local equations $\{\varphi_i\}_{i \in I}$ on an appropriate open covering $\{U_i\}_{i \in I}$ of V. For any point $x \in V_{reg}$, there is some integer $i \in I$ and an open subset $U_x \subset U_i$ containing x such that $\varphi_i = \frac{f_i}{g_i}$ on U_x where f_i and g_i are holomorphic functions on U_x and are relatively prime in the local ring at any point $y \in U_x$. The last statement follows from the fact that the local ring at any regular point of V is a unique factorization domain (see for example, Gunning and Rossi [1], p.72 and p.151). Furthermore, at $x \in U_x$, since $\underline{O}_{V,x}$, the local ring of V at x, is again a unique factorization domain, we can write

$$f_i = \varepsilon_i (f_{i,1})^{m_{i,1}} \cdots\cdots (f_{i,k})^{m_{i,k}}$$
$$g_i = \delta_i (g_{i,1})^{n_{i,1}} \cdots\cdots (g_{i,\ell})^{n_{i,\ell}}$$

where $f_{i,1}, \ldots, f_{i,k}$, $g_{i,1}, \ldots, g_{i,\ell}$ are prime elements in $\underline{O}_{V,x}$, ε_i, δ_i are units in $\underline{O}_{V,x}$ and $i \in I$. Choose U_x so small that the units ε_i and δ_i remain units in U_x. Then, $f_{i,j} = 0$ $g_{i,j'} = 0$ define analytic subsets $C_{j,x}$ and $E_{j',x}$ of codimension one in U_x for $i \in I$, $j = 1, \ldots, k$ and $j' = 1, \ldots, \ell$, respectively ; and hence, a formal sum

$$E_x = \sum_{j=1}^{k} m_{i,j} C_{j,x} - \sum_{j'=1}^{\ell} n_{i,j'} E_{j',x} .$$

can be produced.

In this way, at any point x of V_{reg}, we shall find an open set U_x and a formal sum E_x of analytic subsets of codimension one in U_x. We cover V_{reg} by the open sets U_x, $x \in V_{reg}$, and let φ_x denote the local equation of D on U_x, i.e. the restriction of φ_i to U_x. Since $\frac{\varphi_x}{\varphi_{x'}}$ is a nowhere vanishing holomorphic function on $U_x \cap U_{x'}$, we can patch the analytic subsets $C_{j,x}$ and $E_{j',x}$ together on V_{reg} obtaining analytic subsets $C_1, C_2, \ldots, C_s$ and $E_1, \ldots, E_t$ of V_{reg}. These analytic subspaces are connected in U_x ,

$x \in V_{reg}$, by definition. Therefore, they must be irreducible. Moreover, the integers $m_{i,j}$ and $n_{i,j}$, which appear in the expression of E_x are independent of x, because there is unique factorization in the local rings at points belonging to V_{reg}. It is possible, then, to determine a formal sum

$$D' = \sum_{i=1}^{s} m_i C_i - \sum_{j=1}^{t} n_i E_i \quad .$$

Note that the addition of Cartier divisors is reflected in the addition of these formal sums. If $\bar{C}_i$ and $\bar{E}_j$ are the closure of C_i and E_j in V, then $\bar{C}_i$ and $\bar{E}_j$ are irreducible subvarieties of codimension one in V. Moreover, by our construction, none of them are contained in the singular locus of V. We define h by

$$h(D) = \sum_{i=1}^{s} m_i \bar{C}_i - \sum_{j=1}^{t} n_j \bar{E}_j \quad .$$

2) h(D) determines D at points x of V_{reg} and, thus, determines D unless D has in its support a component of codimension one in V of the singular locus of V. But, as we indicated in a remark just before the proof of this theorem, the codimension of the singular locus on a normal variety is strictly bigger than one. Hence h is injective if V is normal.

3) Suppose that $\sum_{i=1}^{N} n_i C_i$ is a Weil divisor on a smooth variety V. As V is smooth and C_i is of codimension one, for any point x of V, there is an open subset U_x of V containing x such that, in U_x, C_i is defined by an equation

$$f_i = 0 \ .$$

We set $\varphi_x = \prod_{i=1}^{N} (f_i)^{n_i}$. Then, $\{\varphi_x\}_{x \in V}$ determines an element D of $H^0(V, \underline{M}_V^* / \underline{O}_V^*)$ and the mapping

$$\sum_{i=1}^{N} n_i C_i \longrightarrow \{\varphi_x\}_{x \in V}$$

gives the inverse to h. Q.E.D.

Remark 4.13.1. In the proof of Theorem 4.13.3), we have only used the fact that the local ring $\underline{O}_{V,x}$ of V at x is a unique factorization domain. Hence the mapping h is bijective if the local ring $\underline{O}_{V,x}$ of V at any point x is a unique factorization domain.

Let D be a Cartier divisor on a complex variety V and let $\{\varphi_i\}$ be local equations of D with respect to an open covering $\{U_i\}_{i \in I}$ of V. The complex line bundle $[D]$ associated with the divisor D has transition functions $\{g_{ij}\}$ with respect to this covering, where $g_{ij} = \varphi_i / \varphi_j$ on $U_i \cap U_j$. Note that there is a canonical isomorphism between $\underline{O}_V(D)$ and $\underline{O}([D])$. Let ψ be a global section of the line bundle $[D]$. ψ is represented by the collection $\{\psi_i\}$ of holomorphic functions, where ψ_i is holomorphic on U_i and on $U_i \cap U_j$, we have

$$\psi_i = g_{ij} \psi_j .$$

Hence, on $U_i \cap U_j$, we have

$$\frac{\psi_i}{\varphi_i} = \frac{\psi_j}{\varphi_j} .$$

This implies that $\{\frac{\psi_i}{\varphi_i}\}$ defines a global meromorphic function f on V. Moreover the meromorphic function f has the property that the Cartier divisor defined by local equations $f \cdot \varphi_i = \psi_i$ is effective or zero, i.e. $(f) + D \geqq 0$.

For a Cartier divisor, we set

$$\mathbb{L}(D) = \{ f \in \mathbb{C}(V) \mid f = 0 \text{ or } (f) + D \geqq 0 \} .$$

Then $\mathbb{L}(D)$ is a vector space over $\mathbb{C}$. By the above argument it is easy to prove the following :

Lemma 4.14. By a $\mathbb{C}$-linear homomorphism

$$\begin{array}{ccc} H^0(V, \underline{O}_V(D)) & \longrightarrow & \mathbb{L}(D) \\ \cup\!\!\!\shortmid & & \cup\!\!\!\shortmid \\ \psi = \{\psi_i\} & \longmapsto & f = \{\frac{\psi_i}{\varphi_i}\} \end{array},$$

these two vector spaces are isomorphic.

Definition 4.15. The set $\{E \mid E$ Cartier devisor, $E \sim D$, $E > 0\}$ is called a complete linear system associated with D and is written as $|D|$. $\dim |D|$ is, by definition, equal to $\dim_{\mathbb{C}} H^0(V, \underline{O}_V(D)) - 1$.

Let ψ be a non-zero element of $H^0(V, \underline{O}_V(D))$. ψ is represented by a collection $\{\psi_i\}_{i\in I}$ of holomorphic functions with respect to an open covering $\{U_i\}_{i\in I}$. Then the collection $\{\psi_i\}_{i\in I}$ defines a Cartier divisor (ψ). If ψ vanishes at some points on V, then the divisor (ψ) is effective and linearly equivalent to D. On $H^0(V, \underline{O}_V(D)) - \{0\}$, we introduce an equivalence relation $\sim$ as follows :

$\varphi_1 \sim \varphi_2$ if there exists a non-zero constant c such that $\varphi_1 = c\,\varphi_2$.

Then $\varphi_1 \sim \varphi_2$ if and only if $(\varphi_1) = (\varphi_2)$.

On $\mathbb{L}(D) - \{0\}$, we introduce a similar equivalence relation $\sim$ as above. Then $f_1 \sim f_2$ if and only if $(f_1) + D = (f_2) + D$. Hence we have the following :

Lemma 4.16. There are isomorphisms

$$\begin{array}{ccc} H^0(V, \underline{O}_V(D)) - \{0\}/\sim & \longrightarrow & |D|, \\ \cup\!\!\!\shortmid & & \cup\!\!\!\shortmid \\ \psi & \longmapsto & (\psi) \end{array}$$

$$\begin{array}{ccc} \mathbb{L}(D) - \{0\}/\sim & \longrightarrow & |D| \\ \cup\!\!\!\shortmid & & \cup\!\!\!\shortmid \\ f & \longmapsto & (f) + D \end{array}.$$

which commute with the isomorphism in Lemma 4.14. Through these isomorphisms $|D|$ can be endowed with the structure of $\mathbb{P}^{\dim |D|}$.

Definition 4.17. A subset L of D, which is isomorphic to

$M - \{0\}/\sim$,where M is a vector subspace of $H^0(V, \underline{O}_V(D))$, is called a linear system contained in $|D|$. dim L is, by definition, equal to $\dim_{\mathbb{C}} M - 1$.

Let $f : V \longrightarrow W$ be a morphism of complex varieties and D be a Cartier divisor on W. Then there exists an open Stein covering $\{W_i\}_{i \in I}$ such that D has a local equation φ_i on W_i. Then the collection $\{\varphi_i \circ f\}$ defines a Cartier divisor f^*D on V. If we set

$$\varphi^*|D| = \{\varphi^* E \mid E \in |D|\} ,$$

then it is easy to see that $\varphi^*|D|$ is a linear system contained in the complete linear system $|\varphi^* D|$ on V. It is possible that $\varphi^*|D|$ may not be a complete linear system on V. If f is surjective, then

$$\dim \varphi^*|D| = \dim |D| .$$

In the remainder of this section we shall assume that the complex variety V is normal. By the above Theorem 4.13, any Cartier divisor can be considered as a Weil divisor. Hence a Cartier divisor $E \in |D|$ is written in the form

$$\sum_{i=1}^{N} n_i E_i ,$$

where E_i is an irreducible subvariety of codimension one in V and n_i is a positive integer.

On a normal variety V, by definition, two Weil divisors D_1 and D_2 are linearly equivalent (we write $D_1 \sim D_2$) if there exists a meromorphic function f such that

$$D_1 = (f) + D_2 ,$$

where we consider a Cartier divisor (f) as a Weil divisor. For a Weil divisor D we can introduce a vector space $\mathbb{L}(D)$ in the same manner as in the case of Cartier divisors. Also we can introduce a complete linear system $|D|$ associated with a Weil divisor in the same manner as in Definition 4.15. Then it is easily seen that the second half of Lemma 4.16 is also valid for a Weil divisor D.

In the following two definitions, a divisor D is a Cartier divisor or a Weil divisor.

Definition 4.18. An effective Weil divisor F is called a fixed component of a complete linear system $|D|$ (resp. a linear system L) if $E > F$ for any divisor $E \in |D|$ (resp. $E \in L$). The maximal (with respect to the order $\geq$) fixed component is called the fixed part of the complete linear system $|D|$ (resp. the linear system L).

Let F be the fixed part of $|D|$ (resp. a linear system L). Then any element $E \in |D|$ (resp. $E \in L$) has the form $E = E_1 + F$, where E_1 is an effective divisor or zero. E_1 is called the variable part of E. The collection of the variable parts of elements of $|D|$ (resp. L) forms a complete linear system $|E_1|$ (resp. a linear system L_1 contained in $|E_1|$).

Definition 4.19. The point $x \in V$ is called a base point of a complete linear system $|D|$ (resp. a linear system L) if x is contained in the supports of all variable parts of the divisors in $|D|$ (resp. L).

The set of all base points of $|D|$ (resp. L) forms an analytic subset. We call it the base locus of $|D|$ (resp. L). As we have assumed that V is normal, the base locus is of codimension at least two. Note that a complete linear system $|D|$ of a Cartier divisor D is free from fixed components and base points if and only if the invertible sheaf $\underline{O}([D])$ is spanned by its global sections as the $\underline{O}_V$-module.

(4.20) Let D be a Cartier divisor on a normal variety V. Then, in Example 2.4.2, we have defined a meromorphic mapping $\Phi_{[D]}$ (we write hereafter Φ_D) associated with a line bundle $[D]$. We shall examine this meromorphic mapping more closely. We set $N = \dim |D|$. Then the meromorphic mapping Φ_D is written as

$$\begin{array}{rcl} \Phi_D : V & \longrightarrow & \mathbb{P}^N \\ \cup & & \cup \\ z & \longmapsto & (\varphi_0(z):\varphi_1(z): \cdots :\varphi_N(z)) , \end{array}$$

where $\{\varphi_0, \varphi_1, \cdots , \varphi_N\}$ is a basis of $H^0(V, \underline{O}_V(D))$. On the other hand, by the isomorphism described in Lemma 4.14, there exists a basis $\{f_0, f_1, \cdots , f_N\}$ of $\mathbb{L}(D)$ and the above Φ_D is also written as

$$\begin{array}{rcl} \Phi_D : V & \longrightarrow & \mathbb{P}^N \\ \cup & & \cup \\ z & \longmapsto & (f_0(z):f_1(z): \cdots :f_N(z)). \end{array}$$

(This meromorphic mapping is, by definition, the meromorphic mapping defined by meromorphic functions $f_1/f_0, f_2/f_0, \cdots , f_N/f_0$ in Example 2.4.1.) Φ_D is often called as the meromorphic mapping associated with a complete linear system $|D|$. If the complete linear system $|D|$ is free from fixed component and base points, then Φ_D is a morphism. More generally we have the following

Lemma 4.20.1. Φ_D is a morphism outside the base locus of $|D|$ and the singular locus of V.

Proof. Let F be the fixed locus of the complete linear system $|D|$ and let S be the singular locus of V. For any point $x \in V - S$, there exists a neighbourhood U_x of $V - S$ such that F is defined by an equation $f_x = 0$, where f_x is holomorphic on U_x (see the proof of Theorem 4.13, 3)). Then, by the definition of the fixed locus, $\tilde{\varphi}_i = \varphi_i/f_x$, $i = 0, 1, \ldots, N$ are holomorphic in a small neighbourhood of x in $V - S$. Moreover if x is not a base point of $|D|$, then there exists at least one i such that $\tilde{\varphi}_i$ does not vanish at x. In U_x, Φ_D coincides with a meromorphic mapping

$$\begin{array}{ccc} U_x & \longrightarrow & \mathbb{P}^N \\ \Psi & & \Psi \\ z & \longmapsto & (\tilde{\varphi}_0(z):\tilde{\varphi}_1(z): \cdots :\tilde{\varphi}_N(z)) \end{array} .$$

Hence Φ_D is holomorphic at x. Q.E.D.

The above proof shows that if the fixed part of $|D|$ is a Cartier divisor and if $|D|$ has no base locus, then Φ_D is a morphism.

We set $W = \Phi_D(V)$. Then Φ_D defines a meromorphic mapping of V onto W, which we also write $\Phi_D : V \longrightarrow W$. Let G be a graph of this meromorphic mapping and let $\pi : \tilde{V} \longrightarrow G$ be a desingularization of G. We set $p = p_V \circ \pi: \tilde{V} \longrightarrow V$. If B is the base locus of $|D|$ and S is the singular locus of V, then we can assume that π induces an isomorphism between $\tilde{V} - p^{-1}(B \cup S)$ and $G - p_V^{-1}(B \cup S)$. Moreover, by Lemma 4.20.1, $G - p_V^{-1}(B \cup S)$ and $V - B \cup S$ are isomorphic through p_V. We set $\mathcal{E} = p^{-1}(B \cup S)$. Then the morphism $\Psi = p_W \circ \pi : \tilde{V} \longrightarrow W$ is equal to a meromorphic mapping

$$\begin{array}{ccc} \tilde{V} & \longrightarrow & W \subset \mathbb{P}^N \\ \Psi & & \Psi \\ \tilde{z} & \longmapsto & (\hat{\varphi}_0(\tilde{z}):\hat{\varphi}_1(\tilde{z}): \cdots :\hat{\varphi}_N(\tilde{z})), \end{array}$$

where $\hat{\varphi}_i = \varphi_i \circ p$. $\hat{\varphi}_i$ is an element of $H^0(\tilde{V}, \underline{O}_{\tilde{V}}(p^*D))$. As this meromorphic mapping is a morphism and $\tilde{V}$ is non-singular, it is easy to see that the linear system $\varphi^*|D|$ has <u>no</u> base points. Let $\tilde{F}$ be the fixed part of a linear system $\varphi^*|D|$ and let F be the fixed part of $|D|$. Then there exists a divisor $\tilde{F}_1$ such that

$$\tilde{F} = \tilde{F}_1 + p^*F.$$

The divisor $\tilde{F}_1$ appears because of the existence of base points of $|D|$. The support of $\tilde{F}_1$ is contained in $\mathcal{E}$. Hence, if E_1 and $\tilde{E}_1$ are the variable parts of $E \in |D|$ and $p^*E \in \varphi^*|D|$, then we have

$$p^*E_1 = \tilde{E}_1 + \tilde{F}_1$$

and there is a one to one correspondence between the variable parts of

$|D|$ and those of $\varphi^*|D|$.

Let H_λ be a Cartier divisor on W cut out by the hyperplane

(4.20.2) $\lambda_0 X_0 + \lambda_1 X_1 + \cdots + \lambda_N X_N = 0$,

in $\mathbb{P}^N$ where $\lambda = (\lambda_0 : \lambda_1 : \cdots : \lambda_N)$. Then it is easy to see that $\Psi^* H_\lambda + \tilde{F} \in \varphi^*|D|$. Conversely for any element $\tilde{E}_1 + \tilde{F} \in \varphi^*|D|$, there exists a Cartier divisor H_λ on W cut our by the hyperplane (4.20.2) such that $\tilde{E}_1 = \Psi^* H_\lambda$. Hence we have proved the following :

Lemma 4.20.3. There is a one to one correspondence between divisors in $|D|$ and hyperplane divisors on W. The correspondence is given in the following way. An element $E_1 + F \in |D|$ corresponds to a hyperplane divisor H_λ if and only if

$$p^* E_1 = \Psi^* H_\lambda + \tilde{F}_1 .$$

We remark that if V is non-singular and $|D|$ has no base points, then we can take $\tilde{V} = V$. Hence the above correspondence is written as

$$E_1 = \Psi^* H_\lambda .$$

After the above observations, it is easy to prove the well-known theorems of Bertini.

Theorem 4.21. Let D be a Cartier divisor on a normal variety V such that $\dim|D| \geqq 1$.

1) (First theorem of Bertini). If $\dim \Phi_D(V) \geqq 2$, then the variable part of a general member of the complete linear system $|D|$ is irreducible. If $\dim \Phi_D(V) = 1$ and $\dim|D| \geqq 2$, then the variable part of a general member of $|D|$ is reducible and is the sum of prime divisors.

2) (Second theorem of Bertini) The singular locus of the variable part of a general member of $|D|$ is contained in the union of the singular locus of V and the base locus of $|D|$.

Remark 4.22. In the first theorem of Bertini, if $\dim|D| = 1$ and

$\dim \Phi_D(V) = 1$, that is, Φ_D is surjective and $\Phi_D(V) = \mathbb{P}^1$, then for the variable part of a general member of $|D|$, both cases are possible.

Proof of Theorem 4.21. We use freely the same notaions as above. Hence $W = \Phi_D(V)$.

1) When $\dim W \geqq 2$, we can choose three hyperplane divisors H_{λ_0}, H_{λ_1}, H_{λ_2} such that meromorphic functions η_1, η_2 on W induced by meromorphic functions $\dfrac{\lambda_{1,0}X_0 + \lambda_{1,1}X_1 + \cdots + \lambda_{1,N}X_N}{\lambda_{0,0}X_0 + \lambda_{0,1}X_1 + \cdots + \lambda_{0,N}X_N}$,

$\dfrac{\lambda_{2,0}X_0 + \lambda_{2,1}X_1 + \cdots + \lambda_{2,N}X_N}{\lambda_{0,0}X_0 + \lambda_{0,1}X_1 + \cdots + \lambda_{0,N}X_N}$ on $\mathbb{P}^N$ are algebrically independent where $\lambda_i = (\lambda_{i,0}, \lambda_{i,1}, \cdots, \lambda_{i,N})$, $i = 0,1,2$. By the meromorphic mapping Φ_D, we can consider η_1 and η_2 as elements of $\mathbb{C}(V)$. Then η_1 and η_2 are algebrically independent in $\mathbb{C}(V)$. Then by Zariski's lemma (see for example, Hodge and Pedoe [1], Chap. X, §13, Theorem 1, p.78), there exists a constant c such that the field $\mathbb{C}(\eta_1 + c\,\eta_2)$ is algebrically closed in $\mathbb{C}(V)$.

Now we define a meromorphic mapping $\Phi_\eta : \tilde{V} \longrightarrow \mathbb{P}^1$ by

$$\begin{array}{ccc} \tilde{V} & \longrightarrow & \mathbb{P}^1 \\ \cup\!\!\!\shortmid & & \cup\!\!\!\shortmid \\ z & \longmapsto & (1 : \eta_1(z) + c\,\eta_2(z)). \end{array}$$

Let G_η be the graph of the meromorphic mapping Φ_η and let $\hat{\pi}: \hat{V} \longrightarrow G_\eta$ be a desingularization of G_η. We set $\hat{\Psi} = p_{\mathbb{P}^1} \circ \hat{\pi}: \hat{V} \longrightarrow G_\eta \longrightarrow \mathbb{P}^1$ and $\hat{p} = p_{\tilde{V}} \circ \hat{\pi} : \hat{V} \longrightarrow G_\eta \longrightarrow \tilde{V}$. Then, as $\mathbb{C}(\eta_1 + c\eta_2)$ is algebrically closed in $\mathbb{C}(V) = \mathbb{C}(\tilde{V})$, by Corollary 1.10, a fibre $\hat{\Psi}^{-1}(a)$ is connected for any $(1 : a) \in \mathbb{P}^1$. $\hat{\Psi}^{-1}(a)$ is non-singular, hence irreducible except a finite number of a's. Let H_{λ_a} be a hyperplane divisor on W defined by the equation

$$(\lambda_{1,0} + c\lambda_{2,0} + a\lambda_{0,0})X_0 + (\lambda_{1,1} + c\lambda_{2,1} + a\lambda_{0,1})X_1 + \cdots$$
$$+ \cdots + (\lambda_{1,N} + c\lambda_{2,N} + a\lambda_{0,N})X_N = 0.$$

Then from the very definition of the meromorphic functions η_1, η_2, we infer readily that $\hat{p}^* \Psi^* H_{\lambda_a} = \hat{\Psi}^{-1}(a)$. Hence the variable component $E_{a,1}$ of the divisor $E_a \in |D|$ which corresponds to the divisor H_{λ_a} is irreducible. Since there are several choices of H_{λ_0}, H_{λ_1}, H_{λ_2}, if $\dim W \geq 2$, then the variable part of a general member of $|D|$ is irreducible.

Next suppose that $\dim W = 1$ and $\dim|D| \geq 2$. Then a general hyperplane divisor H_λ is written as a sum of d distinct points

$$p_1 + p_2 + \cdots + p_d ,$$

where p_i is a point of W and d is the degree of W in $\mathbb{P}^N$. Hence we can write

$$\Psi^* H_\lambda = \Psi^{-1}(p_1) + \Psi^{-1}(p_2) + \cdots + \Psi^{-1}(p_d) .$$

Then $\Psi^{-1}(p_i)$, $i = 1, 2, \cdots, d$ define Cartier divisors on $\tilde{V}$ and they are mutually disjoint. Hence, from the consideration just before Theorem 4.21, the variable part of a general member of $|D|$ is reducible and a sum of at least d distinct divisors. (Note that the above $\Psi^{-1}(p_i)$ may not be irreducible.)

In the case where $\dim W = 1$ and $\dim|D| = 1$, the variable part of a general member of $|D|$ is irreducible if and only if fibres of Ψ are connected. This is equivalent to saying that $\mathbb{C}(W)$ is algebraically closed in $\mathbb{C}(V)$.

2) Now we shall prove the second part. We consider again the morphism $\Psi : \tilde{V} \longrightarrow W \subset \mathbb{P}^N$. Let A_i be an open set defined by $X_i \neq 0$ and let $\xi_i^1, \xi_i^2, \cdots, \xi_i^N$ be global coordinates of A_i such that

$$\xi_i^1 = \frac{X_0}{X_i}, \cdots, \xi_i^i = \frac{X_{i-1}}{X_i}, \xi_i^{i+1} = \frac{X_{i+1}}{X_i}, \cdots, \xi_i^N = \frac{X_N}{X_i}$$

We set $\tilde{V}_i = \Psi^{-1}(A_i)$. Then the morphism Ψ is represented by

$$\xi_i^k = f_k(z), \quad k = 1, 2, \ldots, N ,$$

where $f_k(z)$ is holomorphic on $\tilde{V}_i$. Let $\mathbb{P}^N$ be another complex projective space with homogeneous coordinates $(Y_0 : Y_1 : \cdots : Y_N)$. (We consider it as the dual projective space of the above $\mathbb{P}^N$.) B_i is an open set defined by $Y_i \neq 0$. We set

$$\eta_i^1 = \frac{Y_0}{Y_i}, \quad \ldots , \eta_i^i = \frac{Y_{i-1}}{Y_i}, \eta_i^{i+1} = \frac{Y_{i+1}}{Y_i}, \quad \ldots , \eta_i^N = \frac{Y_N}{Y_i} .$$

$(\eta_i^1, \cdots , \eta_i^N)$ is a system of global coordinates of B_i. Let $\mathcal{D}$ be a subvariety in $\tilde{V} \times \mathbb{P}^N$ defined by the equations

$$F_k(\eta , z) = \sum_{k=1}^{N} \eta_i^k f_k(z) + 1 = 0,$$

in $\tilde{V}_i \times B_i$, $i = 0, 1, 2, \cdots , N$. Let $q : \mathcal{D} \longrightarrow P^N$ be the projection to $\mathbb{P}^N$. It is clear that $\mathcal{D}$ is a Cartier divisor on $\tilde{V} \times \mathbb{P}^N$ and for any point $\lambda = (\lambda_0 : \lambda_1 : \cdots : \lambda_N) \in \mathbb{P}^N$, $q^{-1}(\lambda)$ defines a Cartier divisor on $\tilde{V}$ which is, by our construction of $\mathcal{D}$, equal to a Cartier divisor $\Psi^* H_\lambda$. Hence if $\mathcal{D}$ is non-singular, then by Corollary 1.7, for a general λ, $\Psi^* H_\lambda$ is non-singular. Then, since the modification $p : \tilde{V} \longrightarrow V$ induces an isomorphism between $\tilde{V} - p(B \cup S)$ and $V - B \cup S$, where B is the base locus of $|D|$ and S is the singular locus of V, the variable part of a general member is non-singular outside of the base locus of $|D|$ and the singular locus of V. But as we have

$$\frac{\partial F_i}{\partial \eta_i^k} = f_k(z) ,$$

and for any point $z \in \mathcal{D} \cap \tilde{V}_i$ at least one $f_k(z)$ does not vanish, $\mathcal{D}$ is non-singular. Q.E.D.

<u>Remark 4.23</u>. For a Cartier divisor D on a normal algebraic variety over an algebrically closed field of characteristic $p > 0$, we can extend the first theorem of Bertini in the following form (see

Zariski [4], Theorem 1.6.3, p.30).

If $\dim \Phi_D(V) \geqq 2$, then the variable part of a general member of $|D|$ is irreducible. If $\dim \Phi_D(V) = 1$, then the variable part of a general member of $|D|$ has a form $p^e\Delta$ where Δ is the sum of prime divisors.

The integer e may be positive because of the existence of inseparable extensions. Moreover, by the same reason the second theorem of Bertini is false in positive characteristics. This is one of the difficulties to generalize our classification theory to the case of positive characteristics.

§ 5. D-dimensions and L-dimensions

Let L be a complex line bundle on a complex variety V and D be a Cartier divisor on V.

All the results of this section hold if the symbol D is replaced by the symbol L, the symbol mD is replaced by the symbol $L^{\otimes m}$, m a positive integer.

At the beginning, we shall assume that the variety V is normal. We set

$$\mathbb{N}(D, V) = \{m > 0 \mid \dim_{\mathbb{C}} H^0(V, \underline{O}_V(mD)) \geqq 1\} .$$

Note that $\mathbb{N}(D, V)$ is a semi-group under the addition induced from the addition of integers. If $\mathbb{N}(D, V)$ is non-empty, for any positive integer $m \in \mathbb{N}(D, V)$, we have a meromorphic mapping Φ_{mD}

$$\begin{array}{cccc} \Phi_{mD} : & V & \longrightarrow & \mathbb{P}^N \\ & \cup\!\!\!\!\!\!\text{\scriptsize{w}} & & \cup\!\!\!\!\!\!\text{\scriptsize{w}} \\ & z & \longmapsto & (\varphi_0(z) : \varphi_1(z) : \cdots : \varphi_N(z)), \end{array}$$

where $\{\varphi_0, \varphi_1, \ldots, \varphi_N\}$ is a basis of $H^0(V, \underline{O}_V(mD))$ (see 2.4.2). Φ_{mD} is not necessarily holomorphic ; however, if $\underline{O}_V(mD)$ is spanned by its global sections, then Φ_{mD} is a morphism. In the language of Cartier divisors, if the complete linear system $|mD|$ of a Cartier divisor mD has no base points and no fixed component, then Φ_{mD} is a morphism (see Lemma 4.20.1 and the remark just after this lemma).

Definition 5.1. The D-dimension (Divisor dimension) of a Cartier divisor D on a variety V is the value $\kappa(D, V)$ defined through the following process :

1) If V is a normal variety, then

$$\kappa(D, V) = \begin{cases} -\infty & \text{if } \mathbb{N}(D, V) = \emptyset \\ \max\limits_{m \in \mathbb{N}(D, N)} (\dim \Phi_{mD}(V)), & \text{if } \mathbb{N}(D, V) \neq \emptyset . \end{cases}$$

2) If V is not normal and $\iota : V^* \longrightarrow V$ is the normalization of the variety V, we set

$$\kappa(D, V) = \kappa(\iota^* D, V^*)$$

where $\iota^* D$ denotes the pull back of the Cartier divisor D to V^*. If we consider a line bundle L instead of a Cartier divisor D, then the number $\kappa(L, V)$ is called the L-dimension (line bundle dimension) of the line bundle L.

By definition, we have

$$\kappa(mD, V) = \kappa(D, V)$$

for a positive integer m. On the other hand if $\dim H^0(V, \underline{O}_V(mD)) \geq 2$ for a positive integer m, then $\kappa(D, V) > 0$. We use this fact freely.

Remark 5.2. Suppose that V is not normal and that

$$f : V' \longrightarrow V$$

is a proper modification with V' normal. Then,

$$\kappa(D, V) = \kappa(f^* D, V') .$$

This assertion is a consequence of Lemma 5.3 below.

Lemma 5.3. Let $f : V \longrightarrow W$ be a surjective morphism of complex varieties. Suppose that W is normal and f is a modification or more generally, every fibre of f is connected. Then, for any Cartier divisor D on W, we have a natural $\mathbb{C}$-linear isomorphism

$$\begin{array}{ccc} f^* : H^0(W, \underline{O}_W(D)) & \longrightarrow & H^0(V, \underline{O}_V(f^* D)) \\ \cup\!\!\!\mathsf{I} & & \cup\!\!\!\mathsf{I} \\ \varphi & \longmapsto & \varphi \circ f \end{array} .$$

Proof. For a locally free sheaf $\underline{F}$ on W we have a canonical isomorphism

$$f_* f^* \underline{F} \xrightarrow{\sim} \underline{F} \underset{\underline{O}_W}{\otimes} f_* \underline{O}_V .$$

In our case, by Proposition 1.13 and Corollary 1.14 we have

$$f_* \underline{O}_V = \underline{O}_W .$$

We set $\underline{F} = \underline{O}_W(D)$. Then there is a canonical isomorphism

$$f_* f^* \underline{O}_W(D) \xrightarrow{\sim} \underline{O}_W(D) .$$

On the other hand there is a canonical isomorphism

$$H^0(W, f_* f^* \underline{O}_W(D)) \xrightarrow{\sim} H^0(V, f^* \underline{O}_W(D)).$$

As $f^* \underline{O}_W(D)$ is isomorphic to $\underline{O}_V(f^* D)$, we have the desired result.

Q.E.D.

Proof of Remark 5.2. Let $\iota : V^* \longrightarrow V$ be the normalization of V. Then V' and V^* are bimeromorphically equivalent. Let $g : V' \longrightarrow V^*$ be a bimeromorphic mapping such that $f = g \circ \iota$. Let $\pi : V^\# \longrightarrow G$ be a desingularization of the graph G of g. We set $h_1 = p_{V'} \circ \pi$, $h_2 = p_{V^*} \circ \pi$, where $p_{V'}$ and p_{V^*} are projections of G onto V' and V^*, respectively. h_1 and h_2 are proper modifications. By our definition, it is easy to see that $f \circ h_1 = \iota \circ h_2$. Then, by the very definition of $\kappa(D, V)$ and by Lemma 5.3, we have equalities

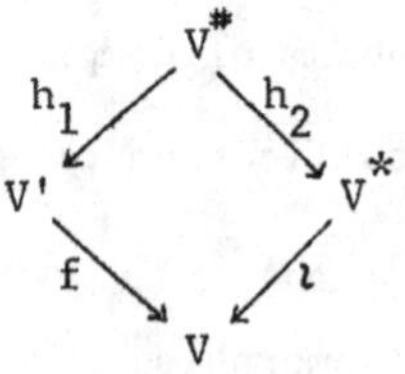

$$\kappa(D, V) = \kappa(\iota^* D, V^*),$$
$$\kappa(\iota^* D, V^*) = \kappa(h_2^* \iota^* D, V^\#) = \kappa(h_1^* f^* D, V^\#),$$
$$\kappa(f^* D, V') = \kappa(h_1^* f^* D, V^\#).$$

Q.E.D.

Example 5.4. 1) Let D be an ample Cartier divisor on a variety V. Then $\kappa(D, V) = \dim V$. (Note that if $\iota : V^* \longrightarrow V$ is a normalization, then $\iota^* D$ is ample if D is ample.)

2) If $-D$ is effective, then $\kappa(D, V) = -\infty$.

3) If mD is a trivial Cartier divisor for a positive integer m,

then $\kappa(D, V) = 0$. More generally, let L be a line bundle with $c_1(L) = 0$ on a Kähler manifold V. Then $\kappa(L, V) \leq 0$. Moreover, $\kappa(L, V) = 0$ if and only if there exists a positive integer m such that $L^{\otimes m}$ is trivial.

4) Let V and W be complex varieties and D be a Cartier divisor on V. Then

$$\kappa(D, V) = \kappa(p^* D, V \times W) ,$$

where $p : V \times W \longrightarrow V$ is a projection .

5) Let D be an effective divisor on an abelian variety A. Then $\kappa(D, A) > 0$. $\kappa(D, A) = \dim A$ if and only if D is ample (see Weil [2], Chap. VI, $n^o 5$ and $n^o 10$).

6) The following is the simplest case of the example due to Zariski (see Zariski [5], p.562-p.564.). Let C be a non-singular curve of degree three in $\mathbb{P}^2$. Let $\mathfrak{a}$ be a divisor class cut out on C by a curve of degree four in $\mathbb{P}^2$. As C is an abelian variety, it is easy to show that there exists twelve distinct points $q_1, q_2, \cdots, q_{12}$ on C such that

$$m(q_1 + q_2 + \cdots + q_{12}) \notin m\mathfrak{a}$$

for any positive integer m. Let S be a surface obtained by the blowing-up at these twelve points $q_1, q_2, \ldots, q_{12}$ and let $\bar{C}$ be the strict transform of C (that is, the closure of the inverse image of $C - \{q_1, q_2, \ldots, q_{12}\}$ in S). Let H be a line in $\mathbb{P}^2$ avoiding all q_i's and let $\bar{H}$ be the strict transform of H. Zariski [5] has shown that the complete linear system $|m(\bar{C} + \bar{H})|$ has a fixed locus $\bar{C}$ for any $m \geq 1$ and $|m\bar{C} + (m - 1)\bar{H}|$ has no fixed components and base points. In our case, using Nakai's criterion for ampleness (see Nakai [1]), we can easily show that the divisor $m\bar{C} + (m - 1)\bar{H}$ is

<u>ample</u>. Hence we have

$$\kappa(\bar{C} + \bar{H}, S) = 2.$$

This example also shows that the graded ring

$$R = \bigoplus_{m=0}^{\infty} H^0(S, \underline{O}_V(m(\bar{C} + \bar{H})))$$

is <u>not</u> finitely generated.

<u>Lemma 5.5</u>. $\kappa(D, V) \leqq a(V) \leqq \dim V$.

<u>Proof</u>. Theorem 3.1 tells us that $a(V) \leqq \dim V$; our definitions, then, imply the lemma without difficulty. Q.E.D.

The mappings Φ_{mD} will play an important role in our theory and it is therefore important that we examine them closely. We assume, again, that V is a <u>normal</u> variety. We set

$$W_m = \Phi_{mD}(V)$$

where $m \in \mathbb{N}(D, V)$. We show now that the function fields $\mathbb{C}(W_m)$ of the algebraic variety W_m do not grow infinitely as m tends to infinity.

<u>Lemma 5.6</u>. There exists a positive integer m_0 such that, for any integer m satisfying $m \geqq m_0$ $m \in \mathbb{N}(D, V)$, we have

$$\mathbb{C}(W_m) = \mathbb{C}(W_{m_0}) .$$

That is, W_m is birationally equivalent to W_{m_0} for any $m \geqq m_0$ such that $m \in \mathbb{N}(D, V)$.

<u>Proof</u>. Let $d > 0$ be the greatest common divisor of the integers belonging to $\mathbb{N}(D, V)$. As $\mathbb{N}(D, V)$ is a semigroup, there is a positive integer ℓ_0 such that $\ell \geqq \ell_0$ implies that $\ell d \in \mathbb{N}(D, V)$. We set $k = \ell_0 d$. Suppose that $\{\varphi_0, \varphi_1, \ldots, \varphi_N\}$ is a basis of $H^0(V, \underline{O}_V(nkD))$. In that case,

$$g_1 = \frac{\varphi_1}{\varphi_0}, \quad g_2 = \frac{\varphi_2}{\varphi_0}, \quad \dots, \quad g_N = \frac{\varphi_N}{\varphi_0}$$

are meromorphic functions on V and $\mathbb{C}(W_{nk})$ is the field $\mathbb{C}(g_1, g_2, \dots, g_N)$. Consider a non-zero element ψ of $H^0(V, \underline{O}_V(kD))$. As $\psi\varphi_1, \psi\varphi_2, \dots, \psi\varphi_N$ are elements of $H^0(V, \underline{O}_V((n+1)kD))$

$$g_1 = \frac{\psi\varphi_1}{\psi\varphi_0}, \quad g_2 = \frac{\psi\varphi_2}{\psi\varphi_0}, \quad \dots, \quad g_N = \frac{\psi\varphi_N}{\psi\varphi_0}$$

are elements of $\mathbb{C}(W_{(n+1)k})$. Therefore the inclusions

$$\mathbb{C}(W_{nk}) \subset \mathbb{C}(W_{(n+1)k}) \subset \mathbb{C}(W_{(n+2)k}) \subset \cdots \subset \mathbb{C}(V)$$

of function fields follow. Theorem 3.1 shows that the function field $\mathbb{C}(V)$ of V is finitely generated. Hence any subfield of $\mathbb{C}(V)$ is finitely generated. There must exist, hence, a positive integer n_0 such that

$$\mathbb{C}(W_{sk}) = \mathbb{C}(W_{n_0 k})$$

for any integer s bigger that n_0.

We set $m_0 = (n_0 + 1)k$. For any integer $m \in \mathbb{N}(D, V)$ with the property $m \geqq m_0$, we write $m = n_0 k + m'$. As the integer d divides m and k, and $m' \geqq k$, $m' \in \mathbb{N}(D, V)$. Hence there must be a non-zero element ψ' of $H^0(V, \underline{O}_V(m'D))$ and, using this ψ' in a manner similar to the use of ψ above, we obtain

$$\mathbb{C}(W_{n_0 k}) \subset \mathbb{C}(W_m) .$$

On the other hand, we have

$$\mathbb{C}(W_m) \subset \mathbb{C}(W_{mn_0}) = \mathbb{C}(W_{n_0 k}) .$$

Hence, it follows that

$$\mathbb{C}(W_{n_0 k}) = \mathbb{C}(W_m) . \qquad \text{Q.E.D.}$$

Continuing with the notation employed in Lemma 5.6, we assert :

<u>Proposition 5.7</u>. $\mathbb{C}(W_{m_0})$ is algebraically closed in $\mathbb{C}(V)$.

<u>Proof</u>. Let $\psi \in \mathbb{C}(V)$ be algebraic over $\mathbb{C}(W_{m_0})$. ψ satisfies an equation

$$X^n + a_1 X^{n-1} + \cdots + a_n = 0$$

where $a_i \in \mathbb{C}(W_{m_0})$, $1 \leq i \leq n$. For a sufficiently large positive integer $m \geq m_0$ and $m \in \mathbb{N}(D, V)$, there are elements

$$b, b_1, \ldots, b_n$$

belonging to $H^0(V, \underline{O}_V(mD))$ such that $a_i = \frac{b_i}{b}$, $1 \leq i \leq n$.

The product $\varphi = b\psi$ satisfies the equation

$$Y^n + b_1 Y^{n-1} + bb_2 Y^{n-2} + \cdots + b^{n-1} b_n = 0 .$$

As φ is a meromorphic section of the line bundle $[mD]$ and the variety V is normal, we obtain

$$\varphi \in H^0(V, \underline{O}_V(mD)).$$

Because $\psi = \frac{\varphi}{b} \in \mathbb{C}(W_m) = \mathbb{C}(W_{m_0})$, the proof of the proposition is complete.

Let $f : V \longrightarrow W$ be a meromorphic mapping. We say $g : V' \longrightarrow W$ is <u>a fibre space associated with</u> f if g is a fibre space and there is a commutative diagram

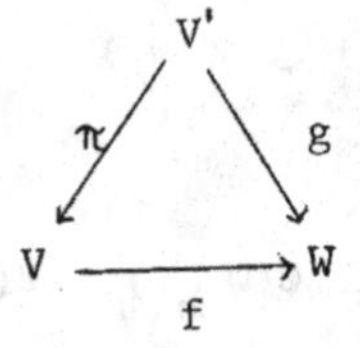

where π is a modification. Let V' and V'' be the graph and a non-singular model of the graph, respectively, of a meromorphic mapping

$\Phi_{mD} : V \longrightarrow W_m$ where V is normal and m is a positive integer. There is, by Lemma 5.3, a commutative diagram

$$\begin{array}{ccc} & \hat{V} & \\ {\scriptstyle\hat{\alpha}}\swarrow & & \searrow{\scriptstyle\Phi_{m\hat{D}}} \\ V & \xrightarrow[\Phi_{mD}]{} & W_m \end{array}$$

where $\hat{V}$ may be V' or V'', $\hat{\alpha}$ is a modification and $\hat{D}$ is the pull back of the Cartier divisor D to $\hat{V}$. In this setting, which we quite often encounter, the next corollary is applicable. Note that $\Phi_{m\hat{D}}$ is a fibre space associated with Φ_{mD}.

Corollary 5.8. If m is an integer such that $m \geqq m_0$, the general fibre of the morphism $\Phi_{m\hat{D}} : \hat{V} \longrightarrow W_m$ is connected. More precisely, there exists an algebraic subset S of W_m such that $W_m - S$ consists of all normal points of W_m and for any $x \in W_m - S$, the fibre $\hat{V}_x$ of $\Phi_{m\hat{D}}$ over x is connected. If W_m is normal, all the fibres of $\Phi_{m\hat{D}}$ are connected.

This is an easy consequence of Corollary 1.10.

Let us assume that

1) V is a complex manifold;

2) $\underline{O}_V(mD)$ is spanned by its global section for some integer m such that $m \geqq m_0$ (m_0 as above) and $m \in \mathbb{N}(D, V)$. That is, $|mD|$ is free from base points and fixed components.

As we have previously indicated, Φ_{mD} is in this case a morphism. Furthermore, we have the following results.

Corollary 5.9. Suppose that V and mD satisfy conditions 1) and 2) above. Then there exists a nowhere dense algebraic subset S of W_m such that

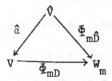

1) for any $w \in W_m - S$, the fibre $V_w = \Phi_{mD}^{-1}(w)$ is a complex manifold;

2) the restriction $[mD]_{|V_w}$ of the line bundle $[mD]$ to V_w is trivial for any $w \in W_m$. Hence, a fortiori, for $w \in W_m - S$, we have

$$\kappa(D_w, V_w) = 0 ,$$

where D_w is the restriction of the Cartier divisor D to V_w.

Proof. By Lemma 4.20.3, any $E \in |mD|$ is written in the form $\Phi_{mD}^{*}H$, where H is a hyperplane divisor on W_m. As we have

$$[mD]_{|V_w} = [E]_{|V_w} = [\Phi_{mD}^{*}H]_{|V_w} = \Phi_{mD}^{*}[H_{|w}] ,$$

$[mD]_{|V_w}$ is trivial. The existence of S is proved in Corollary 1.7. Q.E.D.

The last corollary generalizes to the following theorem of Iitaka which will be of far reaching importance to us.

Theorem 5.10. For a Cartier divisor D on a variety V, we assume that $\kappa(D, V) > 0$. In that case, there is a complex manifold V^* together with a projective manifold W^* and a surjective morphism $f : V^* \longrightarrow W^*$ which have the following properties :

1) a modification $\pi : V^* \longrightarrow V$ exists ;

2) $\dim W^* = \kappa(D, V)$;

3) for a dense subset U of W^* (in the complex topology), each fibre $V_w^* = f^{-1}(w)$, $w \in U$, is irreducible and non-singular ;

4) $\kappa(\pi^* D_w, V_w^*) = 0$ for $w \in U$, where $\pi^* D_w$ denotes the restriction of $\pi^* D$ to V_w^*.

5) If $f^{\#} : V^{\#} \longrightarrow W^{\#}$ is a fibre space satisfying properties 1) through 4) (instead of f), there are bimeromorphic mappings $g : V^* \longrightarrow V^{\#}$ and $h : W^* \longrightarrow W^{\#}$ such that

$$h \circ f = f^{\#} \circ g .$$

$$\begin{array}{ccc} V^* & \xrightarrow{\ g\ } & V^\# \\ {\scriptstyle f}\downarrow & & \downarrow{\scriptstyle f^\#} \\ W^* & \xrightarrow[\ h\ ]{} & W^\# \end{array}$$

Moreover, the fibre space $f : V^* \longrightarrow W^*$ is bimeromorphically equivalent to a fibre space associated with the meromorphic mapping $\Phi_{mD} : V \longrightarrow W_m$ (for instance, the morphism $\Phi_{m\hat{D}}$ appearing in the remark prior to Lemma 5.8 for any integer m such that $m \geqq m_0$ and $m \in \mathbb{N}(L, V)$).

A proof of Theorem 5.10 will be provided in Chapter III §7.

Theorem 5.11. Let $f : V \longrightarrow W$ be a fibre space between non-singular varieties. For any Cartier divisor D on V, there exists an open dense subset U of W such that for any fibre $V_w = f^{-1}(w)$, $w \in U$, the inequality

$$\kappa(D, V) \leqq \kappa(D_w, V_w) + \dim W$$

holds.

Proof. For $m \in \mathbb{N}(D, V)$, we set

$$\mathcal{L}_m = f_*\underline{O}_V(mD) \quad .$$

There exists an open dense subset $U^{(m)}$ of W such that $\mathcal{L}_{m|U^{(m)}}$ is locally free and $V_u = f^{-1}(u)$, $u \in U^{(m)}$, is non-singular (see Theorem 1.4 and Corollary 1.8). Let $g^{(m)} : \mathbb{P}(\mathcal{L}_m) \longrightarrow W$ be the projective fibre space associated with the coherent sheaf $\mathcal{L}_m$ (see (2.9) and (2.10)). The morphism f factors through $g^{(m)}$ and a meromorphic mapping $h^{(m)}$ as the diagram below :

$$\begin{array}{ccc} V & \xrightarrow{\ h^{(m)}\ } & \mathbb{P}(\mathcal{L}_m) \\ & {\scriptstyle f}\searrow \quad \swarrow{\scriptstyle g^{(m)}} & \\ & W & \end{array}$$

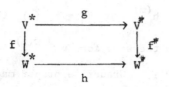

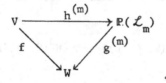

Moreover, if we restrict $h^{(m)}$ to $V_u = f^{-1}(u)$, the restriction $h_u^{(m)} : V_u \longrightarrow \mathbb{P}(\mathcal{L}_m)_u$ is the same for $u \in U^{(m)}$ as the meromorphic mapping Φ_{mD_u} (see (2.10)). Grauert's proper mapping theorem (Theorem 1.4) implies that $\dim_{\mathbb{C}} H^0(V_u, \underline{O}_{V_u}(mD_u))$ is constant for any $u \in U$ where U is an open dense subset of $U^{(m)}$ and that $\dim \Phi_{mD_u}(V_u)$ is constant for any $u \in U$. It follows that

$$\dim h^{(m)}(V) = \dim W + \dim \Phi_{mD_u}(V_u)$$

for any $u \in U$.

On the other hand, as there is a canonical isomorphism

$$H^0(V, \mathcal{L}_m) \xrightarrow{\sim} H^0(V, \underline{O}_V(mD)) ,$$

using a basis of $H^0(V, \underline{O}_V(mD))$, we can construct a meromorphic mapping

$$h : \mathbb{P}(\mathcal{L}_m) \longrightarrow \mathbb{P}^N, \quad N = \dim_{\mathbb{C}} H^0(V, \underline{O}_V(mD)),$$

such that the composition $h \circ h^{(m)} : V \longrightarrow \mathbb{P}^N$ is nothing other than the meromorphic mapping Φ_{mD}. Hence there exists a generically surjective meromorphic mapping $\hat{h} : h^{(m)}(V) \longrightarrow W_m$. Taking m large enough, we obtain the desired inequality :

$$\kappa(D, V) = \dim W_m \leqq \dim h^{(m)}(V) = \dim W + \dim \Phi_{mD_u}(V_u)$$

$$\leqq \dim W + \kappa(D_u, V_u) ,$$

for $u \in U$. Q.E.D.

<u>Lemma 5.12</u>. Let G be a finite group of analytic automorphisms of a normal complex variety V and let $f : V \longrightarrow W = V/G$ be the quotient morphism. For a Cartier divisor D on W, the canonical homomorphism

$$f^* : H^0(W, \underline{O}_W(D)) \longrightarrow H^0(V, \underline{O}_V(f^*D))^G$$

is an isomorphism.

Proof. By Cartan [1], the quotient space W is normal and there is an isomorphism of sheaves

$$\underline{O}_W \xrightarrow{\sim} (f_*\underline{O}_V)^G ,$$

where $(f_*\underline{O}_V)^G$ is the G-invariant subsheaf of $f_*\underline{O}_V$. On the other hand there exist isomorphisms of sheaves

$$f_*\underline{O}_V(f^*D))^G \xrightarrow{\sim} (\underline{O}_W(D) \underset{\underline{O}_W}{\otimes} f_*\underline{O}_V)^G \xrightarrow{\sim} \underline{O}_W(D) \underset{\underline{O}_W}{\otimes} (f_*\underline{O}_V)^G .$$

Hence we have isomorphisms

$$H^0(V, \underline{O}_V(f^*D))^G \xrightarrow{\sim} H^0(W, (f_*\underline{O}_V(f^*D))^G) \xrightarrow{\sim} H^0(W, \underline{O}_W(D)).$$

Q.E.D.

Theorem 5.13. Let $f : V \longrightarrow W$ be a surjective morphism of complex varieties and let D be a Cartier divisor on W. Then we have

$$\kappa(f^*D, V) = \kappa(D, W).$$

Proof. Let $\pi : V^* \longrightarrow V$ be the normalization of V. By definition, we have

$$\kappa(\pi^*f^*D, V^*) = \kappa(f^*D, V).$$

Hence we can assume that V is normal. Let $\iota : W^* \longrightarrow W$ be the normalization of W. Then by the universal property of the normalization, $\iota^{-1}\circ f : V \longrightarrow W^*$ is a morphism. As we have

$$\kappa(\iota^*D, W^*) = \kappa(D, W)$$

and

$$(\iota^{-1}\circ f)^*\iota^*D = f^*D,$$

we can assume that V and W are normal.

As f is surjective, the $\mathbb{C}$-linear map

$$\begin{array}{ccc} f^* : H^0(W, \underline{O}_W(mD)) & \longrightarrow & H^0(V, \underline{O}_V(mf^*D)) \\ \cup\!\!\!\mid & & \cup\!\!\!\mid \\ \varphi & \longmapsto & \varphi\circ f \end{array}$$

is injective. Hence we have an inequality

$$\kappa(f^*D, V) \geq \kappa(D, W) .$$

Let

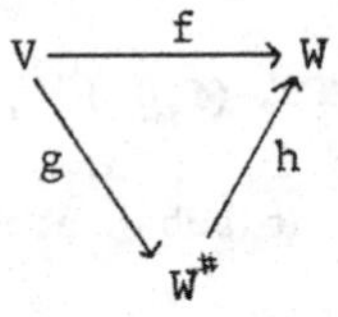

be the Stein factorization of f. As every fibre of g is connected, from Lemma 5.3, we obtain the equality

$$\kappa(g^*h^*D, V) = \kappa(h^*D, W^\#) .$$

(Note that $W^\#$ may not be normal. But using the above argument we can reduce to the case where $W^\#$ is normal. The connectedness of fibres is not changed under such a reduction because of Corollary 1.12) Hence it is enough to show that

$$\kappa(h^*D, W^\#) = \kappa(D, W).$$

Again taking the normalization of $W^\#$, we can assume that $W^\#$ is already normal. Hence it is enough to consider the case where $f : V \longrightarrow W$ is a <u>finite</u> morphism between <u>normal</u> varieties.

Let S be the minimal analytic subset of W such that $f_{|V'} : V' = V - f^{-1}(S) \longrightarrow W' = W - S$ is a finite unramified covering. This unramified covering corresponds to a subgroup H of $\pi_1(W - S)$ of finite index. Then there exists a normal subgroup I of $\pi_1(W - S)$ of finite index such that $I \subset H$. Let $\hat{f} : \hat{V}' \longrightarrow W'$ be a finite Galois covering corresponding to I. Then there exists the finite ramified covering $\hat{f} : \hat{V} \longrightarrow W$ such that $\hat{V}$ is normal and the Galois group $G = \pi_1(W - S)/I$ operates on $\hat{V}$ as a group of analytic automorphisms of $\hat{V}$ (see Grauert and Remmert [1]). Moreover there is a finite morphism $g : \hat{V} \longrightarrow V$ such that $\hat{f} = g \circ f$. As we have already shown that

$$\kappa(\hat{f}^*D, \hat{V}) \geqq \kappa(f^*D, V) \geqq \kappa(D, W),$$

we can assume moreover that $f : V \longrightarrow W$ is a Galois covering of a normal variety W with the Galois group G.

Suppose that $\kappa(f^*D, V) = -\infty$. Then by the above inequality we obtain

$$\kappa(f^*D, V) = \kappa(D, V) = -\infty .$$

Next suppose that $\kappa(f^*D, V) \geqq 0$. Choose a positive integer m such that $\{\varphi_0, \varphi_1, \ldots, \varphi_N\}$ is a basis of $H^0(V, \underline{O}_V(mf^*D))$ and

$$\text{tr. deg}_{\mathbb{C}}\, \mathbb{C}\left(\frac{\varphi_1}{\varphi_0}, \frac{\varphi_2}{\varphi_0}, \ldots, \frac{\varphi_N}{\varphi_0}\right) = \kappa(f^*D, V).$$

We set

$$\prod_{g\in G}\left(X - g^*\left(\frac{\varphi_i}{\varphi_0}\right)\right) = X^n + S_1\left(\frac{\varphi_i}{\varphi_0}\right)X^{n-1} + \cdots + S_n\left(\frac{\varphi_i}{\varphi_0}\right) ,$$

$$i = 1, 2, \ldots, N ,$$

where X is a variable. Then the field $\mathbb{C}\left(\frac{\varphi_1}{\varphi_0}, \frac{\varphi_2}{\varphi_0}, \ldots, \frac{\varphi_N}{\varphi_0}\right)$ is a finite algebraic extension of the field F generated over $\mathbb{C}$ by $S_j\left(\frac{\varphi_i}{\varphi_0}\right)$, $i = 1, 2, \ldots, N$, $j = 1, 2, \ldots, n$. We set

$$\psi = \prod_{g\in G} g^*(\varphi_0) .$$

ψ is not identically zero and is an element of $H^0(V, \underline{O}_V(nmf^*D))^G$. Moreover, we have

$$\psi S_j\left(\frac{\varphi_i}{\varphi_0}\right) \in H^0(V, \underline{O}_V(nmf^*D))^G ,$$

$$i = 1, 2, \ldots, N, \quad j = 1, 2, \ldots, n .$$

Hence, by Lemma 5.12, there exist elements

$\psi_0, \psi_{ij} \in H^0(W, \underline{O}_W(nmD))$ $\quad i = 1, 2, \ldots, N$, $\quad j = 1, 2, \ldots, n$

satisfying

$$f^*(\psi_0) = \psi , \quad f^*(\psi_{ij}) = \psi S_j\left(\frac{\varphi_i}{\varphi_0}\right) .$$

This implies that we can consider the field F as a subfield of $\mathbb{C}(W_{nm})$. Hence we have

$$\kappa(f^*D, V) = \text{tr. deg}_{\mathbb{C}} F \leqq \text{tr. deg}_{\mathbb{C}} \mathbb{C}(W_{nm}) = \kappa(D, W). \qquad \text{Q.E.D.}$$

§6. Kodaira dimension of a complex variety

Let V be an n-dimensional complex manifold. Ω_V^k will denote the sheaf of germs of holomorphic k-forms on V. The canonical line bundle K_V (sometimes we write $K(V)$) of V is, by definition, a complex line bundle whose associated invertible sheaf $\underline{O}(K_V)$ is isomorphic to Ω_V^n. If $\{U_i\}_{i\in I}$ is a small open covering of V for which $(z_i^1, z_i^2, \cdots, z_i^n)$ are local coordinates on U_i, $i \in I$, then the canonical line bundle K_V has transition functions $\{g_{ij}\}_{i,j\in I}$ where

$$g_{ij} = \det\left(\frac{\partial(z_j^1, z_j^2, \cdots, z_j^n)}{\partial(z_i^1, z_i^2, \cdots, z_i^n)}\right) .$$

Also, by these coordinates, any element

$$\varphi \in H^0(V, \underline{O}(mK_V))$$

can be represented by a collection $\{\varphi_i\}_{i\in I}$ where φ_i is a holomorphic function on U_i and, for $i,j \in I$,

$$\varphi_i = \left(\det\left(\frac{\partial(z_j^1, z_j^2, \cdots, z_j^n)}{\partial(z_i^1, z_i^2, \cdots, z_i^n)}\right)\right)^m \varphi_j .$$

The last relation implies that $\varphi_i\cdot(dz_i^1\wedge dz_i^2\wedge\cdots\wedge dz_i^n)^m$ is a well defined global m-tuple n-form. We shall often view an element of $H^0(V, \underline{O}(mK_V))$ as an m-tuple n-form. For convenience, we denote the line bundle $K_V^{\otimes m}$ by mK_V.

Definition 6.1. The m-genus, m a positive integer, of a complex manifold V is

$$P_m(V) = \dim_{\mathbb{C}} H^0(V, \underline{O}(mK_V)) .$$

Definition 6.2. The Kodaira dimension $\kappa(V)$ of a complex manifold V is the L-dimension of the canonical line bundle of V. In other words,

$$\kappa(V) = \kappa(K_V, V) .$$

In order to define the m-genus and the Kodaira dimension of a singular variety, it is necessary to show that the m-genus and the Kodaira dimension are bimeromorphic invariants in the class of complex manifolds. This result is an immediate consequence of the following lemma.

Lemma 6.3. Let $f : V_1 \longrightarrow V_2$ be a generically surjective meromorphic mapping of complex manifolds and $\dim V_1 = \dim V_2$. For each positive integer m, there is a natural $\mathbb{C}$-linear injection

$$f^* : H^0(V_2, \underline{O}(mK_{V_2})) \longrightarrow H^0(V_1, \underline{O}(mK_{V_1})) .$$

Moreover, if f is bimeromorphic, f^* is an isomorphism.

Proof. An application of Theorem 2.5 shows that there exists an analytic subset S of V_1 such that f is holomoprhic on $V_1 - S$ and S is of codimension at least two in V_1. Suppose that ω is an element of $H^0(V_2, \underline{O}(mK_{V_2}))$. Consider coordinate neighbourhoods U_1 and U_2 in V_1 and V_2, respectively, such that $f(U_1 - S) \subset U_2$. We will denote the local coordinates on U_1 and U_2 by $z_1 = (z_1^1, z_1^2, \cdots, z_1^n)$ and $z_2 = (z_2^1, z_2^2, \cdots, z_2^n)$, respectively. ω can be represented on U_2 by the m-tuple n-form

$$\varphi(z_2^1, z_2^2, \cdots, z_2^n) \cdot (dz_2^1 \wedge dz_2^2 \wedge \cdots \wedge dz_2^n)^m .$$

The holomorphic mapping $f : U_1 - S \longrightarrow U_2$ is represented by

$$z_2^i = f^i(z_1^1, \cdots, z_1^n), \ i = 1, 2, \cdots, n ,$$

where f^i is holomorphic on $U_1 - S$.

We specify an element $f^*(\omega)$ of $H^0(U_1 - S, \underline{O}(mK_{V_1}))$ via

$$f^*(\omega) = f(z_1)\cdot\left(\det\left(\frac{\partial(f^1, \cdots, f^n)}{\partial(z_1^1, \cdots, z_1^n)}\right)\right)^m\cdot(dz_1^1\wedge dz_1^2\wedge\cdots\wedge dz_1^n)^m$$

As $S\cap U_1$ is at least of codimension two in U_1 and $\underline{O}(mK_{V_1})$ is an invertible sheaf, by Hartogs' theorem, $f^*(\omega)$ is uniquely extendable to an element in $H^0(U_1, \underline{O}(mK_{V_1}))$. Therefore, since V_1 is covered by open sets of the form U_1 as above, we can associate with every member ω of $H^0(V_2, \underline{O}(mK_{V_2}))$ its pull back $f^*(\omega)$ which belongs to $H^0(V_1, \underline{O}(mK_{V_1}))$. By definition, f^* is linear. As f is generically surjective, f^* is an injection.

Suppose now that $f : V_1 \longrightarrow V_2$ and $g : V_2 \longrightarrow V_3$ are generically surjective meromorphic mappings. Our consideration above implies that $(g\circ f)^* = g^*\circ f^*$. Hence, if f is a bimeromorphic mapping, we have

$$id^* = (f^{-1}\circ f)^* = (f^{-1})^*\circ f^*$$

and

$$id^* = (f\circ f^{-1})^* = f^*\circ (f^{-1})^* .$$

It follows that for a bimeromorphic mapping f, f^* is isomorphic.

Q.E.D.

<u>Corollary 6.4</u>. Let V_1 and V_2 be bimeromorphically equivalent complex manifolds. Then the equalities

$$P_m(V_1) = P_m(V_2)$$

and

$$\kappa(V_1) = \kappa(V_2)$$

hold.

By definition two non-singular models of a complex variety are bimeromorphically equivalent. In combination with Corollary 6.4 this fact shows that the following definition makes sense.

Definition 6.5. Consider a singular variety V and a non-singular model V^* of V. The geometric genus $p_g(V)$, the m-genus $P_m(V)$ and the Kodaira dimension $\kappa(V)$ of V are defined by

$$p_g(V) = p_g(V^*) = P_1(V^*) \ .$$
$$P_m(V) = P_m(V^*) \ .$$
$$\kappa(V) = \kappa(V^*) \ .$$

Example 6.6. 1) Suppose that V is a complex manifold of dimension one. V is thus a compact Riemann surface and therefore, also a complex projective curve. In this case the geometric genus is the genus of a curve. We have the following table.

κ	p_g	P_m $(m\geq 2)$	structure
$-\infty$	0	0	$\mathbb{P}^1$ (rational curve)
0	1	1	elliptic curve
1	$g\geq 2$	$(2m-1)(g-1)$	curves of genus $g\geq 2$

2) An n-dimensional algebraic manifold V is called unirational if there exists a generically surjective rational mapping $f : \mathbb{P}^n \longrightarrow V$. We assert that for a unirational algebraic manifold, we have $P_m(V) = 0$ and $\kappa(V) = -\infty$.

Proof. By virtue of Lemma 6.3, it is enough to consider the case $V = \mathbb{P}^n$. Let H be a hyperplane divisor of $\mathbb{P}^n$. Then $K(\mathbb{P}^n) = [-(n+1)H]$. Hence $mK(\mathbb{P}^n)$ has no holomorphic sections for any positive integer. Q.E.D.

3) If V_1 and V_2 are two complex varieties, we have

$$P_m(V_1 \times V_2) = P_m(V_1) \cdot P_m(V_2),$$

and

$$\kappa(V_1 \times V_2) = \kappa(V_1) + \kappa(V_2).$$

Proof. Clearly, we can assume that V_1 and V_2 are smooth. In that case,

$$K(V_1 \times V_2) = p_1^*(K_{V_1}) \otimes p_2^*(K_{V_2}) ,$$

where $p_i : V_1 \times V_2 \longrightarrow V_i$, $i = 1, 2$ are the natural projections. The Künneth formula (which can be proved by using the theory of harmonic integrals) implies that we have an isomorphism

$$H^0(V_1 \times V_2, \underline{O}(mK(V_1 \times V_2))) \simeq H^0(V_1, \underline{O}(mK_{V_1})) \otimes H^0(V_2, \underline{O}(mK_{V_2})).$$

The proof of 3) is complete.

Remark 6.7. Let V be a complex manifold. For any positive integer m we set

$$P_{-m}(V) = \dim_{\mathbb{C}} H^0(V, \underline{O}(-mK_V)),$$

where $-mK_V$ means the m-th tensor product $(K_V^*)^{\otimes m}$ of the dual bundle K_V^* of K_V. It is easy to show that

$$K(\mathbb{P}^1 \times \mathbb{P}^1) = [\ -2(a \times \mathbb{P}^1 + \mathbb{P}^1 \times b)],$$

where a, b are points on $\mathbb{P}^1$. Using the Künneth formula, we obtain

$$P_{-m}(\mathbb{P}^1 \times \mathbb{P}^1) = (2m + 1)^2 ,$$

for any positive integer m. On the other hand, we have

$$P_{-m}(\mathbb{P}^2) = \binom{3m + 2}{2}$$

for any positive integer m. This shows that $P_{-m}(V)$ is not a bimeromorphic invariant.

Lemma 6.8.1. (Adjunction formula) Let W be an m-dimensional submanifold of an n-dimensional manifold V. The canonical bundle K_W of W has the form

$$K_W = K_V|_W \otimes \overset{n-m}{\wedge} N_W$$

where $K_V|_W$ denotes the restriction of the line bundle K_V to W and N_W is the normal bundle to W in V.

Proof. The normal bundle N_W is the quotient in the exact sequence of vector bundles

$$0 \longrightarrow T_W \longrightarrow T_V|_W \longrightarrow N_W \longrightarrow 0$$

where T_W (resp. T_V) denotes the tangent bundle of W (resp. V) and $T_V|_W$ denotes the restriction of the tangent bundle T_V to W. From this sequence, one derives in turn the following isomorphisms of vector bundles :

$$(\overset{m}{\wedge} T_W) \otimes (\overset{n-m}{\wedge} N_W) \simeq \overset{n}{\wedge}(T_V|_W) ,$$

$$(K_W)^{-1} \otimes (\overset{n-m}{\wedge} N_W) \simeq (K_V)^{-1}|_W ,$$

$$(K_W)^{-1} \simeq (K_V)^{-1}|_W \otimes (\overset{n-m}{\wedge} N_W)^{-1} .$$

Dualizing this isomorphism, the adjunction formula is obtained. Q.E.D.

Corollary 6.8.2. Let $f : V \longrightarrow W$ be a proper morphism of (non compact) complex manifolds. Suppose that any fibre of f is connected and f is of maximal rank at any point of V. Then for any point $w \in W$, we have

$$K(V_w) = K_V|_{V_w}$$

where $V_w = f^{-1}(w)$.

Example 6.9.1. (Kodaira dimension of a complete intersection) An n-dimensional algebraic manifold $V \subset \mathbb{P}^{m+n}$ is called a complete intersection of type $(a_1, a_2, \cdots, a_m)$ if there are m homogeneous equations $F_1, F_2, \cdots, F_m$ of degree $a_1, a_2, \cdots, a_m$ which are defining equations for V in $\mathbb{P}^{m+n}$. We shall prove that if V is a complete intersection of type $(a_1, a_2, \ldots, a_m)$ in $\mathbb{P}^{m+n}$, the canonical bundle K_V of V can be written as

$$K_V = [H]^{\otimes(a_1 + a_2 + \dots + a_m - (m+n+1))}_{|V}$$

where H is a hyperplane in $\mathbb{P}^{m+n}$.

Hence, for such a complete intersection,

$$\kappa(V) = \begin{cases} -\infty, & a_1 + a_2 + \dots + a_m - (m+n+1) < 0, \\ 0, & a_1 + a_2 + \dots + a_m - (m+n+1) = 0, \\ n, & a_1 + a_2 + \dots + a_m - (m+n+1) > 0. \end{cases}$$

Proof. Let D_i be the divisor of $\mathbb{P}^{m+n}$ defined by the equation $F_i = 0$. Clearly, $[D_i] = [H]^{\otimes a_i}$. Because V is a complete intersection of the D_i, $i = 1, 2, \dots, m$, the normal bundle N_V is the direct sum of the normal bundles to the D_i's. Hence we have

$$\bigwedge^m N_V = ([D_1] \otimes [D_2] \otimes \dots \otimes [D_m])|_V .$$

The desired result follows from an application of the adjunction formula. Q.E.D.

Every (non-singular) complete intersection V of dimension $n \geq 2$ is simply connected (see Oka [1] and Mi. Kato [1]). Moreover, the k-th Betti number is given by

$$b_k(V) = \begin{cases} 1 & \text{if } k \equiv 0\ (2),\ 0 \leq k \leq 2n,\ k \neq n, \\ b_n & \text{if } k = n, \\ 0 & \text{otherwise}; \end{cases}$$

where b_n has the following description. The Euler number of a non-singular complete intersection of type $(a_1, a_2, \dots, a_m)$ is equal to the coefficient of z^{n+m} of the power series expansion at the origin of a rational function

$$(1+z)^{n+m+1} \prod_{i=1}^{m} \frac{a_i z}{1+a_i z} .$$

For example, when $n = 1$, we obtain

$$b_1 = a_1 a_2 \cdots a_m (a_1 + a_2 + \ldots + a_m - m - 2) + 2 .$$

This generalizes the usual form of Plücker's formula for a non-singular plane curve.

Proofs of the above statements can be found in L. Fáry [1] and Hirzebruch [1], 22.1, p.159-161.

Example 6.9.2. Let D be a bounded domain in $\mathbb{C}^n$ and let Γ be a discontinuous group of analytic automorphism of D. We assume that Γ operates on D freely and the quotient space D/Γ is compact. Hence $V = D/\Gamma$ is a complex manifold. Then we have

$$\kappa(V) = n.$$

This is proved in Kodaira [1], Theorem 6. Kodaira has shown that the canonical line bundle K_V is ample.

Example 6.9.3. Let $\pi : V \longrightarrow T$ be an analytic fibre bundle over a complex torus T whose fibre and structure group are a complex torus F and the analytic automorphism group $\mathrm{Aut}(F)$ of F. Then we have

$$\kappa(V) \leqq 0 .$$

Proof. As the canonical bundle of a complex torus is trivial, the direct image sheaf $\pi_* \underline{O}_V(mK_V)$ is invertible. Moreover it is easy to see that the line bundle associated with the invertible sheaf is a flat line bundle. Hence we have

$$\dim H^0(V, \underline{O}(mK_V)) = \dim H^0(W, \pi_* \underline{O}_V(mK_V)) \leqq 1.$$

Note that a flat line bundle L on a Kähler manifold has non-zero holomorphic section if and only if L is trivial (see Example 5.4.3)). Hence we conclude that $\kappa(V) = -\infty$ or $\kappa(V) = 0$. Q.E.D.

We shall now restate the result of the last section for D-dimensions in terms of Kodaira dimensions.

Theorem 6.10. 1) Let $f : V \longrightarrow W$ be a generically surjective meromorphic mapping of complex varieties such that $\dim V = \dim W$. Then we have

$$\kappa(V) \geq \kappa(W).$$

2) Let $f : V \longrightarrow W$ be a finite unramified covering of complex varieties. Then we have

$$\kappa(V) = \kappa(W).$$

Proof. 1) We can assume that V and W are smooth. Hence this is an easy consequence of Lemma 6.3.

2) First we assume that W is smooth. Then the canonical bundle K_V is isomorphic to f^*K_W, since f is unramified. Hence, in this case, this is a special case of Theorem 5.13. If W is singular we take a non-singular model W^* of W. Then there exists a finite unramified covering $f^* : V^* \longrightarrow W^*$ such that V^* is a non-singular model of V. This proves the theorem.

Theorem 6.11. Consider an algebraic variety (resp. a complex variety) V of positive Kodaira dimension. There exist a projective manifold (resp. a complex manifold) V^*, a projective manifold W^* and a surjective morphism $f : V^* \longrightarrow W^*$ which satisfy the following conditions :

1) V^* is birationally (resp. bimeromorphically) equivalent to V.

2) $\dim W^* = \kappa(V)$.

3) For a dense subset U of W^* (in the complex topology), each fibre $V^*_w = f^{-1}(w)$, $w \in U$, is irreducible and non-singular.

4) $\kappa(V^*_w) = 0$ for each $w \in U$.

5) If $f^{\#} : V^{\#} \longrightarrow W^{\#}$ is a fibre space satisfying properties 1) through 4) (instead of f), there are birational (resp. bimeromorphic) mappings $g : V^* \longrightarrow V^{\#}$ and $h : W^* \longrightarrow W^{\#}$ such that

$$h \circ f = f^{\#} \circ g .$$

$$\begin{array}{ccc} V^* & \xrightarrow{\ g\ } & V^{\#} \\ f \downarrow & & \downarrow f^{\#} \\ W^* & \xrightarrow[\ h\ ]{} & W^{\#} \end{array} .$$

That is, the fibre space $f : V^* \longrightarrow W^*$ is unique up to birational (resp. bimeromorphic) equivalence.

Moreover, the fibre space $f : V \longrightarrow W$ is birationally (resp. bimeromorphically) equivalent to a fibre space associated to the meromorphic mapping $\Phi_{mK} : V \longrightarrow W_m$ for any integer m such that $m \geqq m_0$ and $m \in \mathbb{N}(K_V, V)$.

In Theorem 6.11, 1), we only assume that V^* is bimeromorphically equivalent to V. Hence Theorem 6.11 is slightly stronger than Theorem 5.10 in its form. Theorem 6.11 is often called the <u>fundamental theorem on the pluricanonical fibrations</u>. Its proof will be given in §7. Unfortunately, at the moment, it is not known whether the above dense set U in 3) is open or not. It is also interesting to know whether all regular fibres of $f : V^* \longrightarrow W$ are of Kodaira dimension zero. The details will be discussed in Remark 7.6, below.

In terms of Kodaira dimensions, Theorem 5.11 becomes the next statement.

<u>Theorem 6.12</u>. Let $f : V \longrightarrow W$ be a fibre space of complex varieties. There exists an open dense set U of W such that for any point $w \in W$, the inequality

$$\kappa(V) \leqq \kappa(V_w) + \dim W$$

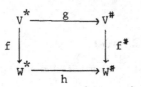

holds where $V_w = f^{-1}(w)$.

Proposition 6.13. Let $f : V \longrightarrow W$ be a fibre space of complex varieties. Suppose that there exists a dense subset U of W such that $\kappa(V_u) = -\infty$ for any $u \in U$, where $V_u = f^{-1}(u)$ is irreducible, then we have

$$\kappa(V) = -\infty .$$

Proof. We can assume that V and W are non-singular. Suppose that $\kappa(V) \geqq 0$. Then there is a non-zero element $\varphi \in H^0(V, \underline{O}(mK_V))$ for a positive integer. Let S be an analytic subset of V defined by the equation

$$\varphi = 0.$$

Let $S_1, S_2, \cdots, S_n$ be irreducible components of S such that

$$f(S_i) \subsetneqq W, \qquad i = 1, 2, \cdots, k ,$$

$$f(S_j) = W, \qquad j = k+1, \cdots, n .$$

We set $\hat{S} = \bigcup_{i=1}^{k} f(S_i)$. $\hat{S}$ is a nowhere dense analytic subset of W. Let T be a nowhere dense analytic subset of W such that f is of maximal rank at any point of $V - f^{-1}(T)$. Then for any point $w \in W - (\hat{S} \cup T)$, φ induces a non-zero element of $H^0(V_w, \underline{O}(mK(V_w)))$ by virtue of Corollary 6.8.2. Hence $\kappa(V_w) \geqq 0$. But as $U \cap (W - (\hat{S} \cup T)) \neq \phi$, this contradicts the assumption that $\kappa(V_u) = -\infty$ for $u \in U$.

Q.E.D.

Chapter III

Fundamental theorems

In this chapter, we shall prove the fundamental theorem on D-dimensions and Kodaira dimensions and two theorems on the asymptotic behaviour of $\ell(mD)$ for a Cartier divisor D on a variety V. These theorems are due to Iitaka [2]. We follow his ideas closely and give the proofs in some detail in order to make clear what can and cannot be proved by these methods.

In §7 we give the proofs of Theorem 5.10 and Theorem 6.11. These theorems are, in some sense, unsatisfactory since we don't know whether the dense set U occurring in both theorems can be chosen as the complement of an algebraic subset (see Remark 7.5 and Remark 7.6).

In §8 we shall study the asymptotic behaviour of $\ell(mD)$ as a function of m. $\ell(mD)$ is not a polynomial in m in general, but Theorem 8.1 and Theorem 8.2 show that it behaves in some ways like a polynomial. These results are deeply related to the structure of the graded ring $R[D] = \bigoplus_{m \geq 0} H^0(V, \underline{O}(mD))$. When V is a surface, the ring $R[D]$ was studied in detail by Zariski (see Zariski [5]). It would be very interesting to determine $R[D]$ for higher dimensional varieties.

§7. Proof of Theorem 5.10 and Theorem 6.11

We begin with the proof of Theorem 5.10. We can assume that the variety V is normal. Moreover, as $\kappa(D, V) > 0$, we can assume that the Cartier divisor is effective. The notations used in this section will be the same as those in §5.

(7.1) There exists a positive integer m_0 such that, if $m \geq m_0$ $m \in \mathbb{N}(D, V)$ and $W_m = \Phi_{mD}(V)$, then $\dim W_m = \kappa(D, V)$ and $\mathbb{C}(W_m)$ is algebraically closed in $\mathbb{C}(V)$ (see Lemma 5.6 and Proposition 5.7). Fix a positive integer $m \geqq m_0$ in $\mathbb{N}(D, V)$ and write W instead of W_m. Let V^* be a non-singular model of the graph G of the meromorphic mapping $\Phi_{mD} : V \longrightarrow W$; and let $b : V^* \longrightarrow V$ be the composite of the projection of G onto V and the desingularization map $V^* \longrightarrow G$. Consider the morphism f defined by the commutative diagram

$$\begin{array}{ccc} & V^* & \\ {}^{b}\swarrow & & \searrow^{f} \\ V & \xrightarrow{\Phi_{mD}} & W \end{array}$$

Since b is a modification and since V^* and V are normal, the $\mathbb{C}$-linear map

$$b^* : H^0(V, \underline{O}_V(\ell D)) \longrightarrow H^0(V^*, \underline{O}_{V^*}(\ell D^*))$$

induced by b is an isomorphism for any positive integer ℓ by Lemma 5.3 where $D^* = b^*(D)$. This implies that the morphism f is nothing other than the mapping Φ_{mD^*}. Hence we can assume that $V^* = V$ and that $\Phi_{mD} = f$, i.e., Φ_{mD} is a morphism and V is non-singular. For simplicity, the symbol $\underline{L}$ will be used instead of $\underline{O}_V(mD)$.

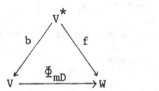

(7.2) We set $N + 1 = \dim_{\mathbb{C}} H^0(V, \underline{L})$. Then $W \subset \mathbb{P}^N$. A hyperplane defined by the equation

$$\lambda_0 X_0 + \lambda_1 X_1 + \cdots + \lambda_N X_N = 0$$

in $\mathbb{P}^N$ cuts off on W a divisor H_λ where $\lambda = (\lambda_0 : \lambda_1 : \cdots : \lambda_N)$. To a divisor H_λ on W there corresponds a divisor $E_\lambda \in |mD|$ (see Lemma 4.20,3). Let F be the fixed part of the complete linear system $|mD|$. We set $E_\lambda^* = f^*(H_\lambda)$. Then $E_\lambda = E_\lambda^* + F$. Furthermore, we set $W_\lambda = W - E_\lambda$.

For an algebraic variety X, by X^s and $\underline{O}_{X^s}$, we shall mean the corresponding $\mathbb{C}$-scheme of finite type over $\mathbb{C}$ and the structure sheaf of that scheme.

Now, by Theorem 1.4, $f_*(\underline{L}^{\otimes n})$ is a coherent $\underline{O}_W$-module for any positive integer n. As W is a projective variety, by Theorem 1.3, 3), there exists an algebraic coherent sheaf $\underline{F}_n$ on W^s such that there is a canonical isomorphism

(7.2.1) $$\underline{F}_n^{an} \xrightarrow{\sim} f_*(\underline{L}\ ) .$$

Since W_λ^s is an affine $\mathbb{C}$-scheme of finite type, $H^0(W_\lambda^s, \underline{F}_n)$ is generated by global sections

$$\psi_0, \psi_1, \cdots, \psi_M$$

as an $H^0(W_\lambda^s, \underline{O}_{W^s})$ module. Using the isomorphism (7.2.1), we consider $\psi_0, \psi_1, \cdots, \psi_M$ as elements of $H^0(W_\lambda, f_*(\underline{L}^{\otimes n})) = H^0(f^{-1}(W_\lambda), \underline{L}^{\otimes n})$.

On the other hand, we have

$$H^0(W_\lambda^s, \underline{F}_n) = H^0(W^s - H_\lambda^s, \underline{F}_n) \subset \bigcup_{e=1}^{\infty} H^0(W^s, \underline{F}_n(eH_\lambda^s))$$

where the sheaf $\underline{F}_n(eH_\lambda^s)$ means the sheaf of germs of local rational sections of $\underline{F}_n$ which have at most a pole of order e on H_λ^s. The sheaf $\underline{F}(eH_\lambda^s)$ is isomorphic to $\underline{F}_n \otimes_{\underline{O}_{W^s}} \underline{O}_{W^s}([eH_\lambda^s])$, hence coherent. By taking e large enough, one can guarantee that $\psi_i \in H^0(S, \underline{F}_n(enH_\lambda^s))$

for $i = 0, 1, \cdots, M$. Again, by Theorem 1.3, 3), we have the isomorphism

$$H^0(W^s, \underline{F}_n(enH_\lambda^s)) \xrightarrow{\sim} H^0(W, f_*(\underline{L}^{\otimes n})(enH_\lambda)).$$

Furthermore,

$$H^0(W, f_*(\underline{L}^{\otimes n})(enH_\lambda)) = H^0(V, \underline{L}^{\otimes n}(enE_\lambda^*)) \subset H^0(V, \underline{L}^{\otimes n}(enE_\lambda))$$

and, hence, we can consider that

$$\psi_0, \psi_1, \cdots, \psi_M \in H^0(V, \underline{L}^{\otimes n}(enE_\lambda)).$$

In this situation, we fix an element $\eta \in H^0(V, \underline{O}_V([en\ H_\lambda]))$ such that $(\eta) = enE_\lambda$. Then $\eta\psi_0, \eta\psi_1, \cdots, \eta\psi_M$ can be considered as elements of $H^0(V, \underline{L}^{\otimes n} \otimes \underline{O}_V([en\ H_\lambda])) = H^0(V, \underline{L}^{\otimes(e+1)n})$.

Let $\psi_{M+1}, \cdots, \psi_{M'}$ be all monomials of degree n in $\varphi_0, \varphi_1, \cdots, \varphi_N$ where $\{\varphi_0, \varphi_1, \cdots, \varphi_N\}$ is a basis of $H^0(V, \underline{L})$. Then $\eta\psi_i \in H^0(V, \underline{L}^{\otimes(e+1)n})$, $i = M+1, \cdots, M'$ and we can define a meromorphic mapping

$$\begin{array}{rccl} h^{(n)} : & V & \longrightarrow & \mathbb{P}^{M'} \\ & \omega & & \omega \\ & z & \longmapsto & (\psi_0(z) : \cdots : \psi_M(z) : \psi_{M+1}(z) : \cdots : \psi_{M'}(z)). \end{array}$$

Let V_n be the projective variety which is the image of $h^{(n)}$. Since $\eta\psi_i \in H^0(V, \underline{L}^{\otimes(e+1)n})$, $i = 0, 1, \cdots, M'$, we infer readily that

$$\mathbb{C}(W) \subset \mathbb{C}(V_n) \subset \mathbb{C}(W_{(e+1)nm})$$

where $W_{(e+1)nm}$ is the projective variety which is the image of $\Phi_{(e+1)nmD}$. As $\mathbb{C}(W_{(e+1)nm}) = \mathbb{C}(W)$, we conclude that there is a commutative diagram

$$\begin{array}{ccc} V & \xrightarrow{h^{(n)}} & V_n \\ & \searrow^{f} & \downarrow g_n \\ & & W \end{array}$$

where g_n is a birational map.

Since $f_*(\underline{L}^{\otimes n})$ is coherent, we can find a proper algebraic subset S_n of W such that $f_*(\underline{L}^{\otimes n})|_{W-S_n}$ is a locally free sheaf over $W - S_n$, $\underline{L}$ is flat over $W - S_n$, and that, by Theorem 1.4,3), there is an isomorphism

$$(7.2.2) \quad f_*(\underline{L}^{\otimes n})_w \xrightarrow{\sim} H^0(V_w, \underline{L}_w^{\otimes n})$$

for any point $w \in W - S_n$, where $V_w = f^{-1}(w)$. If $w \in W_\lambda - S_n$, by the isomorphism (7.2.2), the restriction $\psi_{i,w}$ of ψ_i to a fibre V_w (where ψ_i belongs to $H^0(f^{-1}(W_\lambda), \underline{L}^{\otimes n})$) can be viewed as an element of $H^0(V_w, \underline{L}_w^{\otimes n})$. Corollary 1.8 shows that there is an algebraic subset T of W such that $W - T$ is non-singular and such that each fibre V_w, $w \in W - T$ is non-singular. On the other hand, as W_λ^s is an affine scheme, the algebraic coherent sheaf $\underline{F}_n$ is spanned by its global sections as $\underline{O}_{W_\lambda^s}$- module. Hence by the isomorphisms (7.2.1) and (7.2.2), $H^0(V_w, \underline{L}_w^{\otimes n})$ is spanned by $\psi_{i,w}$, $i = 0, 1, \cdots, M$, for $w \in W_\lambda - S_n$. Hence, for any fibre V_w, $w \in W_\lambda - (S_n \cup T)$, the meromorphic mapping

$$h_w^{(n)} : V_w \longrightarrow \mathbb{P}^{M'} ,$$

which is the restriction of $h^{(n)}$ to V_w, is bimeromorphicallly equivalent to the meromorphic mapping Φ_{mnD_w}. Moreover, we have $h_w^{(n)}(V_w) = V_{n,w} = g_n^{-1}(w)$. As V_n and W are birationally equivalent, there is a nowhere dense algebraic subset B_n of V_n and a nowhere dense algebraic subset A_n of W such that g_n induces an isomorphism between $V_n - B_n$ and $W - A_n$. Hence, for any point $w \in W_\lambda - (S_n \cup T \cup A_n)$, $h_w^{(n)}(V_w) = V_{n,w}$ must be a point. This means that

$$\dim_{\mathbb{C}} H^0(V_w, \underline{O}_{V_w}(mnD_w)) = 1 .$$

As H_λ, T and $S_n \cup A_n$ are nowhere dense subset of W, $Y = H_\lambda \cup T \bigcup_{n=1}^{\infty} (S_n \cup A_n)$ is a Baire set of the first category. As W is a Baire space, $U = W - Y$ is dense in W (see for example Bourbaki [1] §5).

Also, whenever $w \in U$, we have

$$\dim_{\mathbb{C}} H^0(V_w, \underline{O}_{V_w}(nmD_w)) = 1, \quad n = 1, 2, \dots .$$

Since

$$\dim_{\mathbb{C}} H^0(V_w, \underline{O}_{V_w}(nD_w)) \leqq \dim_{\mathbb{C}} H^0(V_w, \underline{O}_{V_w}(nmD_w)) ,$$

for any positive integer m, we conclude that

$$\kappa(D_w, V_w) = 0 ,$$

for $w \in U$. Since for any $w \in U$, W is non-singular at x, V_w is irreducible by Corollary 5.8.

Let $\delta : W^* \longrightarrow W$ be a desingularization of W obtained by finite succession of monoidal transformations as in Theorem 2.12. Then $g = \delta^{-1} \circ f : V \longrightarrow W^*$ is a generically surjective meromorphic mapping. As $W - T$ is non-singular, on $V' = V - f^{-1}(T)$ g is holomorphic.

Let G be the graph of the meromorphic mapping g and let $\pi^* : V^* \longrightarrow G$ be a desingularization of G. Then $\pi = p_V \circ \pi^* : V^* \longrightarrow V$ is a modification and π induces an isomorphism between $\pi^{-1}(V')$ and V. As $W - T$ and $W^* - \delta^{-1}(T)$ are isomorphic, $f^* = f \circ \pi : V^* \longrightarrow W^*$ and $U^* = \delta^{-1}(U)$ satisfy the required conditions 1) 2) 3) and 4) of Theorem 5.10 .

(7.3). Next we shall show that the fibre space $f^* : V^* \longrightarrow W^*$ is unique up to bimeromorphic equivalence.

Let $f^{\#} : V^{\#} \longrightarrow W^{\#}$ be a fibre space of complex manifolds which has the following properties.

(1) There is a modification $\pi^{\#} : V^{*} \longrightarrow V$.

(2) $\dim W^{\#} = \kappa(D, V)$.

(3) For some dense subset $U^{\#}$ of $W^{\#}$, each fibre $V_u^{\#}$ of $f^{\#}$, $u \in U^{\#}$, is irreducible and non-singular.

(4) $\kappa(((\pi^{\#})^{*}D)_u, V_u^{\#}) = 0$ for $u \in U^{\#}$ where $((\pi^{\#})^{*}D)_u$ denote the restriction of $(\pi^{\#})^{*}D$ to $V_u^{\#}$.

As the fibre space $f^{*} : V^{*} \longrightarrow W^{*}$ constructed above is bimeromorphically equivalent to the fibre space $f : V \longrightarrow W$, it is enough to show that $f^{\#} : V^{\#} \longrightarrow W^{\#}$ is bimeromorphically equivalent to $f : V \longrightarrow W$. Moreover we can assume that $V^{\#} = V$ because of the above assumption (1).

Now we want to find a commutative diagram

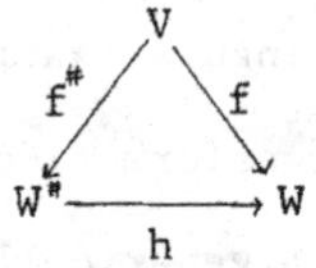

where h is a bimeromorphic mapping.

We fix a basis $\{\varphi_0, \varphi_1, \cdots, \varphi_N\}$ of $H^0(V, \underline{O}_V(mD))$ such that f is defined by this basis. There exists a nowhere dense analytic subset S of $W^{\#}$ such that the sheaf $\underline{L}^{\#} = f_{*}^{\#}\underline{O}_V(mD)$ is locally free on $X = W^{\#} - S$ and f is of maximal rank at any point of $f^{-1}(X)$.

Suppose that for a point $x \in X$, $\varphi_{i,x} = \varphi_i|_{V_x}$, $\varphi_{j,x} = \varphi_j|_{V_x}$ are linearly independent. Then the sheaf $\overset{2}{\wedge}\underline{L}^{\#}$ is locally free on X and $\varphi_i \wedge \varphi_j$ determines a non-zero section of $\overset{2}{\wedge}\underline{L}^{\#}$ over X. Hence there exists an open neighbourhood Y of x in X such that

$\varphi_{i,y}$ and $\varphi_{j,y}$ are linearly independent for any $y \in Y$. This implies that for any $y \in Y$,

$$\kappa(D_y, V_y) > 0 \quad .$$

But as $Y \cap U^{\#} \neq \phi$, this contradicts the assumption (4). Hence, for any point $x \in X$, $\varphi_{i,x}$ and $\varphi_{j,x}$ are always linearly dependent. This implies that $f(f^{\# -1}(x))$ is always a point for any $x \in X$.

On the other hand, from the morphisms $f^{\#}$ and f we can construct a morphism

$$(f^{\#}, f) : V \longrightarrow W^{\#} \times W.$$

Let G be the image of V of the morphism $(f^{\#}, f)$. G is an irreducible analytic subset. As $f(f^{\# -1}(x))$ is a point for any $x \in X$, the morphism $p_{W^{\#}}^{-1}(X) \longrightarrow X$ is one to one where $p_{W^{\#}} : G \longrightarrow W^{\#}$ is the projection. Moreovere, $p_{W^{\#}}^{-1}(X)$ is open dense in G. Hence $p_{W^{\#}} : G \longrightarrow W^{\#}$ is a modification and satisfies the condition (M) in § 2. This implies that there exists a meromorphic mapping $h : W^{\#} \longrightarrow W$ such that $h \circ f^{\#} = f$. As f and $f^{\#}$ are surjective, h is generically surjective. Then by f and h, we obtain inclusions of function fields

$$\mathbb{C}(W) \subset \mathbb{C}(W^{\#}) \subset \mathbb{C}(V).$$

As $\mathbb{C}(W)$ is algebrically closed in $\mathbb{C}(V)$ and $\dim W = \dim W^{\#}$ by the assumption (2), $\mathbb{C}(W) = \mathbb{C}(W^{\#})$. Hence h is bimeromorphic. This completes the proof of the uniqueness of the fibre space up to bimeromorphic equivalence.

(7.4) The proof of Theorem 6.11 is an easy modification of that of Theorem 5.10 by virtue of Lemma 6.3 and Corollary 6.8.2.

Remark 7.5. 1) Let k be the algebraic closure of a finitely generated subfield of $\mathbb{C}$. Let V be an algebraic variety defined over k. Then the above arguments can be carried over the field k and it is not difficult to show that we can choose a dense set U in W^* in such a way that all points of W which are generic points over k in the sense of Weil [1] are contained in U. Hence if V is an algebraic variety defined over k, we can say that for any generic fibre V_w^* of f^* over k, we have $\kappa(D_w^*, V_w^*) = 0$, where $D_w^* = (\pi^* D)_w$. This statement is also true for a generic fibre in the sense of Grothendieck.

2) We use the same notations as those in (7.1) and (7.2). Let φ be a non-zero element of $H^0(V, \underline{L}^{\otimes n})$ and let $\tilde{T}$ be an analytic set of V defined by the equation

$$\varphi = 0 .$$

We decompose $\tilde{T}$ into irreducible analytic sets $T_1, T_2, \cdots, T_\ell$ in such a way that

$$f(T_i) \subsetneqq W, \quad i = 1, 2, \cdots, k,$$

$$f(T_j) = W, \quad j = k+1, \cdots, \ell .$$

We set $\hat{T} = \bigcup_{i=1}^{k} f(T_i)$. Then for any point $x \in W' = W - (\hat{T} \cup S_n \cup T)$, the restriction $\varphi_x = \varphi|_{V_x}$ of φ to the fibre V_x is a non-zero element of $H^0(V_x, \underline{L}_x)$.

Hence we have

$$\kappa(D_x, V_x) \geqq 0,$$

for any $x \in W'$. It is very plausible that the equality holds. At the moment, however, we don't know even whether a much weaker statement that we can choose the above dense set U as an open set does hold or not.

Remark 7.6. In this remark we shall discuss about Theorem 6.11. We have already remarked above that it is not known whether the dense set U can be chosen as an open set or not. This is true if the following conjecture is true.

Conjecture. The Kodaira dimension κ is upper semi-continuous under small deformations of complex manifolds.

The Kodaira dimension and pluri-genera are not invariant under small deformations. (Nakamura [1]. See §17, below). But for surfaces, Iitaka [1], II has shown that these are invariant under (global) deformations. Hence we have the following.

Corollary 7.7. If $c\kappa(V) = \dim V - \kappa(V) \leq 2$, or the pluri-canonical system $|m K_V|$ has no base points and no fixed components for some positive integer m, then for any regular fibre V_w^* of f^* in Theorem 6.11, we have

$$\kappa(V_w^*) = 0 ,$$

where by definition, the fibre V_w^* is regular if f^* is of maximal rank at any point of V_w^*.

It seems very likely that the Corollary 7.7 is true in general.

§ 8. Asymptotic behaviour of $\ell(mD)$

In this section we shall formulate the theorems in terms of Cartier devisors. But if we read D as L, a line bundle, the theorems hold without modifications for any line bundle L.

Let D be a Cartier divisor on a variety V and let $\iota : V^* \longrightarrow V$ be the normalization of V. We define $\ell(mD) = \dim_{\mathbb{C}} H^0(V^*, \underline{O}_{V^*}(m\iota^* D))$.

In this section we shall prove the following two theorems. In these theorems, the integer d denotes the largest common divisor of $\mathbb{N}(D, V)$.

Theorem 8.1. There exist positive numbers α, β and a positive integer m_0 such that the following inequalities hold for any integer $m \geqq m_0$:

$$\alpha\, m^{\kappa(D,\ V)} \leqq \ell(mdD) \leqq \beta\, m^{\kappa(D,\ V)} .$$

Theorem 8.2. Suppose that $\kappa(D, V) > 0$. Then for any positive integer p, there exist a positive number γ and a positive integer m_1 such that the inequality

$$\ell(mdD) - \ell((m-p)dD) \leqq \gamma\, m^{\kappa(D,\ V)-1} ,$$

holds for any positive integer $m \geqq m_1$.

Remark 8.3. Iitaka has used the inequality in Theorem 8.1 to define D-dimensions and Kodaira dimensions of varieties (see Iitaka [2]). Therefore, his difinition and our definition are equivalent.

Remark 8.4. The existence and nature of the limit $\lim\limits_{m\in\mathbb{N}(V)} \dfrac{P_m(V)}{m^{\kappa(V)}}$ is a matter of great interest.

For example, if V is a non-singular complete curve of gemes $g \geqq 2$,

$$\lim \frac{P_m(V)}{m} = 2(g-1) \quad .$$

If V is a non-singular surface of general type we have

$$\lim \frac{P_m(V)}{m^2} = \frac{1}{2} c_1^2$$

where c_1 is the first Chern class of the minimal model of V. More generally if the ring $R(V)=\bigoplus_{m\geqq 0} H^0(V, \underline{O}(mK))$ is finitely generated, the limit exists.

When V is a non-singular surface, the ring R(V) is finitely generated. (See Mumford [4], where the case $\kappa(V) = 2$ is treated. For a surface of $\kappa(V) = 1$, this is an easy consequence of the canonical bundle formula for elliptic surface due to Kodaira. See Kodaira [3], I, Theorem 12, p.772.) But it is not known whether for a complex manifold V of dimension $\geqq 3$, $R(V)$ is finite generated or not. On the other hand, if D is a Cartier divisor on an algebraic manifold, it may happen that the ring $\bigoplus_{m\geqq 0} H^0(V, \underline{O}(mD))$ is not finitely generated (see Example 5.4, 6)).

Proof of Theorem 8.1.

(8.5). We can assume that V is normal. There exists a positive integer n_0 such that, for any integer $n \geqq n_0$, we have $nd \in \mathbb{N}(D, V)$ where d is the largest common divisor of $\mathbb{N}(D, V)$ (see the proof of Lemma 5.6).

Suppose that Theorem 8.1 is valid for the effective Cartier divisor E linearly equivalent to ndD, $n \geqq n_0$. We have

$$\alpha\, m^{\kappa(E,\, V)} \leqq \ell(mE) \leqq \beta\, m^{\kappa(E,\, V)} \quad .$$

Note that

$$\kappa(E, V) = \kappa(D, V).$$

We write

$$m = m_1 n + m_2 , \quad 0 \leqq m_2 < n .$$

It follows that

$$\ell(mdD) \leqq \ell((m_1 + 1)ndD) = \ell((m_1 + 1)E)$$
$$\leqq \beta(m_1 + 1)^{\kappa(D, V)} \leqq \beta(\frac{m}{n} + 1)^{\kappa(D, V)} \leqq \beta' \cdot m^{\kappa(D, V)} ,$$

for a positive number β'. Similary there are inequalities

$$\ell(mdD) \geqq \ell((m_1 - 1)ndD) = \ell((m_1 - 1)E)$$
$$\geqq \alpha(m_1 - 1)^{\kappa(D, V)} \geqq \alpha(\frac{m}{n} - 1)^{\kappa(D, V)} \geqq \alpha' \cdot m^{\kappa(D, V)},$$

for a positive number α'. Thus, Theorem 8.1 is valid for the divisor D. Hence we may replace D by E which is linearly equivalent to ndD, $n \geqq n_0$.

Suppose hence, that D is an effective Cartier divisor on a normal variety V. In that case, $\mathbb{N}(D, V) = \mathbb{N}$ and $d = 1$. Moreover, as a result of the above calculations, we can assume that $\mathbb{C}(W_1)$ is algebrically closed in $\mathbb{C}(V)$ and $\dim W_1 = \kappa(D, V)$, where $W_1 = \Phi_D(V)$ (see Lemma 5.6 and Lemma 5.7).

Lemma 8.6. Let E be an effective Cartier divisor on an n-dimensional projective variety V. Then there is a positive number β such that

$$\ell(mE) \leqq \beta m^n , \quad m \geqq 1 .$$

Moreover, if E is ample, there exists a positive number α such that

$$\alpha m^n \leqq \ell(mE) , \quad m \geqq 1 .$$

Proof. If E is ample, $\ell(mE)$ is a polynomial in m of degree n for sufficiently large m. Hence there are positive numbers α, β such that

$$\alpha\, m^n \leqq \ell(mE) \leqq \beta\, m^n .$$

If E is not ample, there exists an effective divisor defined by a hypersurface section of sufficiently high degree such that $E + H$ is ample. It follows that, for a suitable positive number β' the inequalities

$$\ell(mD) \leqq \ell(m(E + H)) \leqq \beta' \cdot m^n$$

hold. Q.E.D.

Now we introduce the notations $W = W_1'$, $f = \Phi_D : V \longrightarrow W$, $\underline{L} = \underline{O}_V(D)$. Moreover by the same reason as in (7.1), we can take V to be non-singular and f to be a morphism.

We set $N + 1 = \dim_{\mathbb{C}} H^0(V, \underline{L})$. Then $W \subset \mathbb{P}^N$.

A hyperplane difined by

$$\lambda_0 X_0 + \lambda_1 X_1 + \cdots + \lambda_N X_N = 0$$

in $\mathbb{P}^N$ cuts out on W a positive divisor H_λ, $\lambda = (\lambda_0 : \lambda_1 : \cdots : \lambda_N)$ on W. To a positive divisor H_λ on W, there corresponds a divisor $E_\lambda \in |D|$. We set $E_\lambda = E_\lambda^* + F$, where F is the fixed part of the complete linear system $|D|$ and $E_\lambda^* = f^*(H_\lambda)$.

By the above lemma 8.6 there exists a positive number α such that

$$\ell(mD) = \ell(mD_\lambda) \geqq \ell(mE_\lambda^*) \geqq \dim H^0(W, \underline{O}_W(mH_\lambda))$$

$$\geqq \alpha\, m^{\kappa(D,\, V)} ,$$

since $\kappa(D, V) = \dim W$. This proves the first inequality in Theorem 8.1.

Next, we shall show that the other inequality holds. We shall write the divisor F as a sum $F = \sum n_\gamma A_\gamma$ where A_γ's are the irreducible components of F.

We set

$$L = \sum_{f(A_\gamma)=W} n_\gamma A_\gamma , \quad F^* = \sum_{f(A_\gamma)\neq W} n_\gamma A_\gamma .$$

Then for any positive integer m, we have

$$|mD| \ni mL + mF^* + mE_\lambda^* .$$

Let $\sum \ell_j B_j \in |mD|$ be a general member of $|mD|$ where B_j's are the irreducible components.

We set

$$L_m = \sum_{f(B_j)=W} \ell_j B_j , \quad F_m^* = \sum_{f(B_j)\neq W} \ell_j B_j .$$

If Y is a subset of W determined as in (7.2), then for any point $w \in W - Y$, we have

$$\dim_{\mathbf{C}} H^0(V_w, \underline{O}_{V_w}(mD_w)) = 1 .$$

On the other hand, for any point $w \in W - \{Y \cup f(\operatorname{suppt} F^*) \cup f(\operatorname{suppt} F_m^*)\}$ we have isomorphisms

$$\underline{O}_{V_w}(mD_w) \simeq \underline{O}_{V_w}(L_m \mid V_w) \simeq \underline{O}_{V_w}(mL \mid V_w) .$$

It follows that

$$\dim H^0(V_w, \underline{O}_{V_w}(L_m|V_w)) = \dim H^0(V_w, \underline{O}_{V_w}(mL|V_w)) = 1.$$

Moreover on V_w, the divisors $L_m|V_w$ and $mL|V_w$ are linearly equivalent, since

$$L_m + F_m^* \sim mL + mF^* + mE_\lambda^* .$$

Hence we have

$$mL|V_w = L_m|V_w .$$

Because we can choose w to be any point of dense subset

$$W - \{Y \cup f(\operatorname{suppt} F^*) \cup f(\operatorname{suppt} F_m^*)\}$$

in W and f maps any component of L_m and mL onto W, we conclude that $mL = L_m$. Therefore $L_m = mL$ is contained in the fixed part of $|mD|$. Hence we have the equalities

$$\ell(mD) = \ell(L_m + F_m^*) = \ell(F_m^*) = \ell(mF^* + mE_\lambda^*) .$$

As f maps every component of F^* onto a lower dimensional subvariety of W, there exists an effective Cartier divisor H on W such that $F^* \leqq f^*(H) = H^*$. Lemma 8.6 implies, hence, that there exists a positive number β with the property

$$\begin{aligned}\ell(mF^* + mE_\lambda^*) \leqq \ell(m(H^* + E_\lambda^*)) &= \dim H^0(W, \underline{O}_W(m(H + E_\lambda))) \\ &\leqq \beta m^{\kappa(D, V)} .\end{aligned}$$

This completes the proof of Theorem 8.1.

<u>Proof of Theorem 8.2</u>.

(8.7) Arguments similar to those above show that if the theorem is true for an effective Cartier divisor linearly equivalent to kD for a positive integer k, then the theorem is also true for a Cartier divisor D.

Thus we can assume that D is an effective Cartier divisor, $\mathbb{C}(W)$ is algebrically closed in $\mathbb{C}(V)$ and $\dim W = \kappa(D, V)$, where $W = \Phi_D(V)$.

In addition, we can suppose that $f = \Phi_D : V \longrightarrow W$ is a morphism, V is non-singular and W is normal. We employ the same notations as in the proof of Theorem 8.1. We have already shown that any member E_λ of the complete linear system $|D|$ has a form

$$E_\lambda = E_\lambda^* + L + F^*, \quad E_\lambda^* = f^*(H_\lambda)$$

where H_λ is a hyperplane divisor of W and $L + F^*$ is the fixed part of $|D|$ such that f maps any irreducible component of L onto W and f maps any irreducible component of F^* onto a lower dimensional subvariety of W. Moreover, we have shown that mL is contained in the fixed part of the complete linear system $|mD|$. Therefore, the equalities

$$\ell(mD) = \ell(mE_\lambda^* + mF^*) ,$$

$$\ell((m - p)D) = \ell((m - p)E_\lambda^* + (m - p)F^*)$$

hold. Let G be an effective Cartier divisor on W such that

$$F^* < f^*(G) .$$

Using Lemma 5.3 and Corollary 5.8, we have

$$\ell(mE_\lambda^* + mF^*) \leqq \ell(mE_\lambda^* + mf^*G) = \dim H^0(W, \underline{O}_W(m(H_\lambda + G))),$$

$$\begin{aligned} \ell((m - p)D) &= \ell(mE_\lambda^* + mf^*(G) - (pE_\lambda^* + pF^* + mf^*(G))), \\ &\geqq \ell(mE_\lambda^* + mf^*(G) - (pE_\lambda^* + (p + m)f^*(G)) \\ &= \dim H^0(W, \underline{O}_W(m(H_\lambda + G) - \{pH_\lambda + (p + m)G\})) . \end{aligned}$$

When H^* is a very ample divisor of W such that $\bar{H} = H^* + pH_\lambda + (p + m)G$ is very ample, there exists a prime divisor H which is lineary equivalent to $\bar{H}$. Then we have

$$\begin{aligned} &\dim H^0(W, \underline{O}_W(m(H_\lambda + G) - \{pH_\lambda + (p + m)G\})) \\ &\geqq \dim H^0(W, \underline{O}_W(m(H_\lambda + G) - \bar{H})) \\ &= \dim H^0(W, \underline{O}_W(m(H_\lambda + G) - H)). \end{aligned}$$

Hence we have

$$\ell(mD) - \ell((m - p)D) \leqq \dim H^0(W, \underline{O}_W(m(H_\lambda + G)) - \dim H^0(W, \underline{O}_W(m(H_\lambda + G) - H)).$$

By the exact sequence of sheaves

$$0 \longrightarrow \underline{O}_W(m(H_\lambda + G) - H) \longrightarrow \underline{O}_W(m(H_\lambda + G)) \longrightarrow \underline{O}_H(m(H_\lambda + G)_{|H}) \longrightarrow 0,$$

we obtain

$$\begin{aligned} &\dim H^0(W, \underline{O}_W(m(H_\lambda + G))) - \dim H^0(W, \underline{O}_W(m(H_\lambda + G) - H)) \\ &\leqq \dim H^0(H, \underline{O}_H(m(H_\lambda + G)_{|H})) . \end{aligned}$$

By Lemma 8.6, there exists a positive number γ such that

$$\dim H^0(H, \underline{O}(m(H_\lambda + G)\,|_H)) \leqq \gamma\, m^{\kappa(D, V)-1}$$

for any positive integer m, since $\dim H = \dim W - 1 = \kappa(D, V) - 1$. This completes the proof of Theorem 8.2.

<u>Remark 8.8</u>. Using Lemma 6.3, we can easily show that for a variety V, Theorem 8.1 and Theorem 8.2 hold if we replace $\ell(mdD)$ and $\kappa(D, V)$ by P_{md} and $\kappa(V)$, respectively.

Chapter IV

Classification of algebraic varieties and complex varieties

For any complex variety, we have introduced two structures of fibre spaces, that is, the fibre space induced by the algebraic reduction (Definition 3.3) and the fibre space associated with the pluricanonical mappings (Theorem 6.11). In this chapter, we shall introduce another fibre space, the fibre space associated with the Albanese mapping (see Definition 9.17 below). Using these three fibre spaces, we shall discuss several problems on classification theory.

In §9, first we shall recall the definition and elementary properties of the Albanese mappings. Then the fibre space associated with the Albanese mapping will be introduced. Finally we shall define certain bimeromorphic invariants of complex varieties which play an important role in our classification theory.

To study the fibre space associated with Albanese mapping more closely, it is important to know the structure of the variety which is the image of the Albanese mapping. For that purpose, in §10, we shall study the structure of subvarieties in a complex torus. It turns out that such a subvariety has a simple structure (see Thorem 10.3 and Theorem 10.9). As a corollary, we obtain the important fact that the Albanese mapping is surjective if and only if the Kodaira dimension of its image is zero (see Corollary 10.6). This section is taken from Ueno [3].

In §11, we shall give an outline of our classification theory and discuss related conjectures and problems. By Theorem 6.11, the classification theory is reduced to

1) the study of complex varieties of hyperbolic, parabolic and elliptic type (see Definition 11.1 below);

2) the study of fibre spaces whose general fibre is a complex variety of parabolic type.

It will be shown that the following Conjecture C_n is the central problem of the <u>classification of algebraic varieties</u>.

<u>Conjecture C_n</u>. Let $f : V \longrightarrow W$ be a fibre space of an n-dimensional algebraic manifold V over an algebraic manifold W. Then we have

$$\kappa(V) \geqq \kappa(W) + \kappa(V_w) \quad ,$$

where $V_w = f^{-1}(w)$, $w \in W$ is a general fibre of f.

Up to now, we have a few affirmative answers to this conjecture (see Theorem 11.5.3 and Theorem 11.5.4). It is urgent to give good sufficient conditions under which Conjecture C_n holds.

§9. Albanese mappings and certain bimeromorphic invariants

In this section we recall the definition of Albanese tori and Albanese mappings and collect elementary properties of Albanese mappings. For that purpose first we shall consider cohomology groups of complex manifolds.

Let V be a compact complelx manifold of dimension n and let Ω_V^p be the sheaf of germs of holomorphic p-forms on V. Then there is an exact sequence

$$0 \longrightarrow \mathbb{C} \longrightarrow \mathcal{O}_V \xrightarrow{d} \Omega_V^1 \xrightarrow{d} \Omega_V^2 \longrightarrow \cdots \xrightarrow{d} \Omega_V^n \longrightarrow 0 ,$$

where d is the exterior differentiation (see, for example, Kodaira and Morrow [1] chap.2, Th. 6.4, p.81).

Viewing this exact sequence as a complex, we can construct a Cartan-Eilenberg resolution of this complex by flabby sheaves :

$$\begin{array}{ccccccccccccc}
 & & 0 & & 0 & & 0 & & 0 & & & 0 & \\
 & & \downarrow & & \downarrow & & \downarrow & & \downarrow & & & \downarrow & \\
0 & \longrightarrow & \mathbb{C} & \longrightarrow & \mathcal{O}_V & \xrightarrow{d} & \Omega_V^1 & \xrightarrow{d} & \Omega_V^2 & \xrightarrow{d} \cdots & \xrightarrow{d} & \Omega_V^n & \longrightarrow 0 \\
 & & \downarrow & & \downarrow & & \downarrow & & \downarrow & & & \downarrow & \\
0 & \longrightarrow & F^0 & \longrightarrow & F^{0,0} & \xrightarrow{d^{(1,0)}} & F^{1,0} & \xrightarrow{d^{(2,0)}} & F^{2,0} & \xrightarrow{d^{(3,0)}} \cdots & \xrightarrow{d^{(n,0)}} & F^{n,0} & \longrightarrow 0 \\
 & & \downarrow{\scriptstyle d^{(1)}} & & \downarrow{\scriptstyle d^{(0,1)}} & & \downarrow{\scriptstyle d^{(1.1)}} & & \downarrow{\scriptstyle d^{(2,1)}} & & & \downarrow{\scriptstyle d_1} & \\
0 & \longrightarrow & F^1 & \longrightarrow & F^{0,1} & \xrightarrow{d^{(1,1)}} & F^{1,1} & \xrightarrow{d^{(2,1)}} & F^{2,1} & \xrightarrow{d^{(3,1)}} \cdots & \xrightarrow{d^{(n,1)}} & F^{n,1} & \longrightarrow 0 \; . \\
 & & \downarrow{\scriptstyle d^{(2)}} & & \downarrow{\scriptstyle d^{(0,2)}} & & \downarrow{\scriptstyle d^{(1,2)}} & & \downarrow{\scriptstyle d^{(2,2)}} & & & \downarrow{\scriptstyle d_2} &
\end{array}$$

Hence, taking the global sections of the sheaves $F^{p,q}$, we obtain a double complex $\{K^{p,q}\}$, $K^{p,q} = H^0(V, F^{p,q})$. From the double complex $\{K^{p,q}\}$ we have two spectral sequences

$${}'E_1^{p,q} = H^q(V, \Omega^p) \Longrightarrow \mathbb{H}^{p+q} ,$$

$$''E_1^{p,q} = \text{Ker } d^{(p,q+1)}/\text{Im } d^{(p,q)} \Longrightarrow \mathbb{H}^{p+q} .$$

As the exact sequence

$$0 \longrightarrow F^p \longrightarrow F^{0,p} \longrightarrow F^{1,p} \longrightarrow F^{2,p} \longrightarrow \cdots$$

is a flabby resolution of the flabby sheaf F^p, we have

$$''E_1^{p,q} = \begin{cases} H^0(V, F^p) & q = 0 \\ 0 & q \neq 0. \end{cases}$$

Hence we have

$$''E_2^{p,q} = \begin{cases} H^p(V, \mathbb{C}) & q = 0 \\ 0 & q \neq 0. \end{cases}$$

Therefore the second spectral sequence degenerates and it follows that

$$\mathbb{H}^k \xrightarrow{\sim} H^k(V, \mathbb{C}) .$$

From the first spectral sequence we derive, thus, a spectral sequence

$$E_1^{p,q} = H^q(V, \Omega^p) \Longrightarrow H^{p+q}(V, \mathbb{C}) ,$$

which we call the <u>Hodge spectral sequence</u>.

The theory of harmonic integrals says the following :

<u>Theorem 9.1</u>. (Hodge, Kodaira, de Rham) If V is a Kähler manifold, then

1) the Hodge spectral sequence

$$H^q(V, \Omega^p) \Longrightarrow H^{p+q}(V, \mathbb{C})$$

degenerates ;

2) there exists a $\mathbb{C}$-anti-linear isomorphism between $H^q(V, \Omega^p)$ and $H^p(V, \Omega^q)$.

<u>Proof</u>. By the definition of a spectral sequence, the Hodge spectral sequence degenerates if and only if

$$\sum_{p+q=n} \dim_{\mathbb{C}} H^q(V, \Omega^p) = \dim_{\mathbb{C}} H^n(V, \mathbb{C}).$$

If V is Kähler, this is the case (see Kodaira and Morrow [1], Th. 5.4, p.114, Hodge [1], Weil [2] .)

The second part of the theorem comes from the fact that on a Kähler manifold the Laplacian Δ and the complex conjugate operator "$-$" commute (see Kodaira and Morrow [1] Th. 5.3, p.113, Weil [2].)

Q.E.D.

Remark 9.2. We set

$$h^{p,q} = \dim_{\mathbb{C}} H^q(V, \Omega^p).$$

1) From the definition of spectral sequences we infer that, even if V is not Kähler, we have

$$c_n(V) = \sum (-1)^{p+q} h^{p,q} ,$$

where $c_n(V)$ is the Euler number of V, i.e.,

$$c_n(V) = \sum_{k=0}^{2n} (-1)^k \dim_{\mathbb{C}} H^k(V, \mathbb{C}) .$$

2) By the Serre duality, we have

$$h^{p,q} = h^{n-p,n-q} ,$$

(see Serre [1], Kodaira and Morrow [1], Th.4.2, p.104).

3) By the above theorem, if V is Kähler, we have

$$h^{p,q} = h^{q,p} ,$$

$$b_k(V) = \sum_{p+q=k} h^{p,q} ,$$

where $b_k(V)$ is the k-th Betti number of V. Hence, if k is odd, $b_k(V)$ is even. Moreover, we know that $b_{2\ell}(V) \geqq 1$ for $0 \leqq \ell \leqq n$, (see Kodaira and Morrow [1], Corollary 1, p.114).

Deligne [1] has shown that the above theorem is also true if V is a compact algebraic manifold. Note that a compact algebraic manifold is projective if and only if V is Kähler (see Moishezon [1]). Here we shall prove the following :

Corollary 9.3. Suppose that there exists a surjective morphism $f : M \longrightarrow V$ of an n-dimensional compact Kähler manifold M onto an n-dimensional compact complex manifold V. Then the statements 1) and 2) in Theorem 9.1 are also valid for V.

Proof. First we shall prove that via f^*, $H^q(V, \Omega_V^p)$ is mapped injectively into $H^q(M, \Omega_M^p)$.

We consider $H^q(V, \Omega_V^p)$ as the Dolbeault cohomology group. Assume that a type (p, q)-form $\omega \in H^q(V, \Omega_V^p)$ is mapped to zero. Then for any type $(n-p, n-q)$-form $\omega' \in H^{n-q}(V, \Omega_V^{n-p})$, we have

$$0 = \int_M f^*(\omega) \wedge f^*(\omega') = \int_M f^*(\omega \wedge \omega')$$

$$= \deg(f) \int_V \omega \wedge \omega' ,$$

where $\deg(f)$ is the number of points in $f^{-1}(a)$ for a general point $a \in V$. Hence, by the Serre duality, we conclude that $\omega = 0$.

A similar argument shows that $H^k(V, \mathbb{C})$ is mapped injectively via f^*.

Thus, there is an injection of the Hodge spectral sequence of V into that of M ;

$$\begin{array}{ccc} H^q(V, \Omega_V^p) & \Longrightarrow & H^{p+q}(V, \mathbb{C}) \\ \downarrow & & \downarrow \\ H^q(M, \Omega_M^p) & \Longrightarrow & H^{p+q}(M, \mathbb{C}) . \end{array}$$

Moreover, this injection induces the commutative diagram

$$\begin{array}{ccccc}
H^q(V, \Omega_V^{p-1}) & \xrightarrow{d_1^{p-1,q}(V)} & H^q(V, \Omega_V^p) & \xrightarrow{d_1^{p,q}(V)} & H^q(V, \Omega_V^{p+1}) \longrightarrow \\
\downarrow & \circlearrowright & \downarrow & \circlearrowright & \downarrow \qquad \circlearrowright \\
H^q(M, \Omega_M^{p-1}) & \xrightarrow{d_1^{p-1,q}(M)} & H^q(M, \Omega_M^p) & \xrightarrow{d_1^{p,q}(M)} & H^q(M, \Omega_M^{p+1}) \longrightarrow .
\end{array}$$

As the spectral sequence

$$H^q(M, \Omega_M^p) \Longrightarrow H^{p+q}(M, \mathbb{C}),$$

degenerates, $d_1^{p,q}(M) = 0$. Hence $d_1^{p,q}(V) = 0$ and the spectral sequence

$$H^q(V, \Omega_V^p) \Longrightarrow H^{p+q}(V, \mathbb{C})$$

also degenerates. This proves the assertion 1).

Hence we have

$$H^k(V, \mathbb{C}) = \bigoplus_{p+q=k} H^q(V, \Omega_V^p) .$$

As $H^k(V, \mathbb{C})$ and $\overline{H^k(V, \mathbb{C})}$ (the complex conjugate) are $\mathbb{C}$-anti-isomorphic, we have

$$H^k(V, \mathbb{C}) = \bigoplus_{p+q=k} \overline{H^q(V, \Omega_V^p)} .$$

In addition, for a type (p, q)-form $\omega \in H^q(V, \Omega_V^p)$, we have

$$f^*(\bar{\omega}) = \overline{f^*(\omega)} .$$

This implies that

$$f^*(\overline{H^q(V, \Omega_V^p)}) = \overline{f^*(H^q(V, \Omega_V^p))} \subset H^p(M, \Omega_V^q) ,$$

and

$$f^*(\overline{H^q(V, \Omega_V^p)}) \cap H^{p'}(V, \Omega_V^{q'})) = 0 \quad \text{if} \quad (p', q') \neq (p, q) .$$

As f^* is injective, it follows that

$$\overline{H^q(V, \Omega_V^p)} = H^p(V, \Omega_V^q) . \qquad \text{Q.E.D.}$$

Remark 9.4. 1) If V is an algebraic manifold, there exists a projective manifold $\hat{V}$ and a birational surjective morphism

$\hat{V} \longrightarrow V$ (Chow's lemma, EGA II, 5.6.2, Mumford [1], Chap. I §10) More generally, if V is a Moishezon manifold (i.e. $a(V) = \dim V$), then such a projective manifold $\hat{V}$ exists (Moishezon[1]). In these cases, V satisfies the above condition given in Corollary 9.3.

2) If V is an analytic surface (i.e. a compact complex manifold of dimension 2), the Hodge spectral sequence degenerates. This is an easy consequence of Kodaira [3], I, Theorem 3, p.755. But assertion 2) in Theorem 9.1 is not necessarily valid for surfaces. For example, a Hopf surface has numerical invariants $h^{1,0} = 0$, $h^{0,1} = 1$ (see §18).

<u>Corollary 9.5</u>. If the Hodge spectral sequence of V degenerates, then any holomorphic form on V is d-closed.

<u>Proof</u>. For the spectral sequence

$$H^q(V, \Omega_V^p) \Longrightarrow H^{p+q}(V, \mathbb{C}),$$

$d_1^{0,p} : H^0(V, \Omega_V^p) \longrightarrow H^0(V, \Omega_V^{p+1})$ is nothing other than the usual exterior differentiation d. As the spectral sequence degenerates, $d_1^{0,p} = d$ is a zero map. Q.E.D.

Next we recall the definition of the Albanese torus.

<u>Definition 9.6</u>. $(A(V), \alpha)$ is called an Albanese torus of a compact complex manifold V if it satisfies the following conditions :

1) $A(V)$ is a complex torus and $\alpha : V \longrightarrow A(V)$ is a morphism;

2) for a morphism $g : V \longrightarrow T$ of V into a complex torus T, there exists a unique Lie group homomorphism $h : A(V) \longrightarrow T$ and a unique element a of T such that

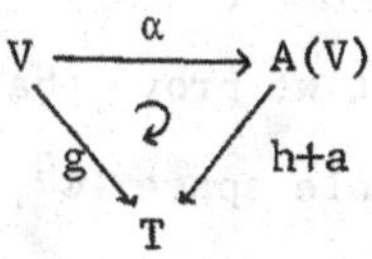

$$g(x) = h(\alpha(x)) + a\ , \quad x \in V.$$

By definition, if (A_1, α_1) and (A_2, α_2) are Albanese tori of V, then A_1 and A_2 are analytically isomorphic as Lie groups and there exists a uniquely determined element $a_1 \in A_1$ such that

$$\alpha_1(x) = \alpha_2(x) + a_1 , \quad x \in V.$$

Thus an Albanese variety of V is, if it exists, uniquely determined up to translations and we can say "the" Albanese torus $(A(V), \alpha)$ of V. We often call the complex torus $A(V)$ itself the Albanese torus of V and the morphism $\alpha : V \longrightarrow A(V)$ the Albanese mapping.

Theorem 9.7. For any compact complex manifold V, there exists the Albanese torus $(A(V), \alpha)$.

Proof. Here we give the proof due to Blanchard [1].

Let $\gamma_1, \gamma_2, \ldots, \gamma_{b_1}$ be a basis of the free part of $H_1(V, \mathbb{Z})$ and let $\{\omega_1, \omega_2, \ldots, \omega_q\}$ be a maximal set of linearly independent d-closed holomorphic 1-forms on V.

$$\mathbb{Z} \cdot \begin{pmatrix} \int_{\gamma_1} \omega_1 \\ \int_{\gamma_1} \omega_2 \\ \vdots \\ \int_{\gamma_1} \omega_q \end{pmatrix} + \mathbb{Z} \cdot \begin{pmatrix} \int_{\gamma_2} \omega_1 \\ \int_{\gamma_2} \omega_2 \\ \vdots \\ \int_{\gamma_2} \omega_q \end{pmatrix} + \cdots + \mathbb{Z} \cdot \begin{pmatrix} \int_{\gamma_{b_1}} \omega_1 \\ \int_{\gamma_{b_1}} \omega_2 \\ \vdots \\ \int_{\gamma_{b_1}} \omega_q \end{pmatrix}$$

form a subgroup Δ of $\mathbb{C}^q$.

First we prove that the vector subspace generated by Δ in $\mathbb{C}^q$ is the whole space $\mathbb{C}^q$. Assume the contrary. Then there is a non-zero d-closed holomorphic form ω such that

$$\int_{\gamma_i} \omega = 0 \ , \quad i = 1, 2, \ldots, b_1 \ .$$

This implies that ω defines the zero cohomology class in $H^1(V, \mathbb{C})$. Therefore, there exists a differentiable function f on V such that $\omega = df$.

As ω is a holomorphic 1-form, we can choose f to be a holomorphic function. Hence f is constant and $\omega = 0$. This is a contradiction.

Let $\bar{\Delta}$ be the smallest <u>closed</u> Lie subgroup of $\mathbb{C}^q$ containing Δ such that the connected component of $\bar{\Delta}$ containing 0 is a vector subspace of $\mathbb{C}^q$. Then $A(V) = \mathbb{C}^q/\bar{\Delta}$ is a complex torus. Let $p : \mathbb{C}^q \longrightarrow A(V)$ be the natural map. Now we fix a point $x_0 \in V$ and define a morphism $\alpha : V \longrightarrow A(V)$ by

$$\begin{array}{ccc} V & \longrightarrow & A(V) \\ \cup\!\!\!\!\!\shortmid & & \cup\!\!\!\!\!\shortmid \\ x & \longmapsto & p(v(x)), \end{array}$$

where

$$v(x) = \begin{pmatrix} \int_{x_0}^{x} \omega_1 \\ \int_{x_0}^{x} \omega_2 \\ \vdots \\ \int_{x_0}^{x} \omega_q \end{pmatrix} .$$

Next we show that $(A(V), \alpha)$ is the Albanese torus of V.

Let $g : V \longrightarrow T$ be a holomorphic mapping of V into a complex torus T. We can assume that $g(x_0) = 0$, $T = \mathbb{C}^n/L$ where L is a lattice of $\mathbb{C}^n$ and $(w_1, \ldots, w_n)$ are coordinates of the vector space $\mathbb{C}^n$. Let $\tilde{V}$ be a universal covering manifold of V. Then the morphism $g : V \longrightarrow T$ can be lifted to a morphism $\tilde{g} : \tilde{V} \longrightarrow \mathbb{C}^n$.

The morphism $\tilde{g}$ is represented by n holomorphic functions g_i, $i = 1, 2, \cdots, n$ on $\hat{V}$ such that $w_i = g_i(\tilde{z})$. Moreover, g_i satisfies the following equality

$$g_i(\gamma(\tilde{z})) = g_i(\tilde{z}) + a_i(\gamma), \quad \gamma \in \pi_1(V, x_0), \tag{9.7.1}$$

where $\begin{pmatrix} a_1(\gamma) \\ a_2(\gamma) \\ \vdots \\ a_n(\gamma) \end{pmatrix} \in L$ and we consider $\pi_I(V, x_0)$ as a covering transformation group of $\tilde{V}$.

As dg_i is invariant under the action of $\pi_1(V, x_0)$, dg_i induces a d-closed holomorphic 1-form on V. Thus there exists an $n \times q$ matrix M such that

$$\begin{pmatrix} dg_1 \\ dg_2 \\ \vdots \\ dg_n \end{pmatrix} = M \cdot \begin{pmatrix} \omega_1 \\ \omega_2 \\ \vdots \\ \omega_q \end{pmatrix}. \tag{9.7.2}$$

This relation and (9.7.1) imply that, for any element $e \in \Delta$, $M \cdot e \in L$. By the definition of $\bar{\Delta}$, Δ is dense in $\bar{\Delta}$. Hence for any element $\bar{e} \in \bar{\Delta}$, $M \cdot \bar{e} \in L$. This implies that the $\mathbb{C}$-linear homomorphism $\tilde{h} : \mathbb{C}^q \longrightarrow \mathbb{C}^n$ defined by the left multiplication by the matrix M induces a complex Lie group homomorphism $h : A(V) \longrightarrow T$. From the relations (9.7.1) and (9.7.2) we infer readily that

$$g = h \circ \alpha .$$

The uniqueness of h is clear from the construction. Q.E.D.

<u>Corollary 9.8</u>. 1) We have

$$\dim A(V) \leqq \dim H^0(V, d\underline{O}_V),$$

where $H^0(V, d\underline{O}_V)$ denotes the vector space consisting of all d-closed holomorphic 1-forms.

2) If V satisfies the assertions 1) and 2) of Theorem 9.1, then

$$\dim A(V) = h^{1,0} = h^{0,1} .$$

Proof. The first assertion is clear from the construction of the Albanese torus. If V satisfies the assertions 1) and 2) of Theorem 9.1, then every holomorphic 1-form is d-closed and $h^{1,0} = \frac{1}{2} b_1$. Furthermore, it is easy to show that

$$\det \begin{pmatrix} M \\ \overline{M} \end{pmatrix} \neq 0,$$

where

$$M = \begin{pmatrix} \int_{\gamma_1} \omega_1 , & \int_{\gamma_2} \omega_1 , & \cdots , & \int_{\gamma_{b_1}} \omega_1 \\ \int_{\gamma_1} \omega_2 , & \int_{\gamma_2} \omega_2 , & \cdots , & \int_{\gamma_{b_1}} \omega_2 \\ \cdots & \cdots & \cdots & \cdots \\ \int_{\gamma_1} \omega_q , & \int_{\gamma_2} \omega_q , & \cdots , & \int_{\gamma_{b_1}} \omega_q \end{pmatrix} ,$$

and $\{\omega_1, \omega_2, \cdots, \omega_q\}$ is a basis of $H^0(V, \Omega^1_V)$. Hence the lattice Δ in the above proof of Theorem 9.7 is of rank 2q. Thus, $\overline{\Delta} = \Delta$.

Q.E.D.

The following example is due to Blanchard [1].

Example 9.9. Let T be a 2-dimensional complex torus whose period matrix is represented by $\begin{pmatrix} 2\pi\sqrt{-1} & a & b & c \\ 0 & 1 & \sqrt{-1} & g \end{pmatrix}$ with respect to global coordinates (x, y) of T (i.e. the coordinates of the universal covering $\mathbb{C}^2$ of T) where

$$\mathrm{Re}(c) \neq \mathrm{Re}(a) \cdot \mathrm{Re}(g) + \mathrm{Re}(b) \cdot \mathrm{Im}(g) .$$

We consider two copies of $\mathbb{C} \times T$ whose coordinates are written as

(z_1, x_1, y_1) and (z_2, x_2, y_2), respectively, where the global coordinates (x_i, y_i) are the same as the global coordinates (x, y) of T. Patching together two $\mathbb{C}\times T$'s via the relation

$$z_2 = \frac{1}{z_1}, \quad y_2 = y_1, \quad x_2 = x_1 + \log z_1 ,$$

we obtain a three-dimensional compact complex manifold V. Then the projection of $\mathbb{C}\times T$ onto the first factor $\mathbb{C}$ induces a surjective morphism $f : V \longrightarrow \mathbb{P}^1$ such that V is a torus bundle over $\mathbb{P}^1$ whose fibre is T by this morphism f. It is easily shown that $\omega = dy_1 = dy_2$ is the only linearly independent d-closed holomorphic 1-form on V. If we take g in such a way that Re(g) and Im(g) are irrational numbers, then Δ in the proof of Theorem 9.7 is not a closed subgroup of $\mathbb{C}$. Hence $\overline{\Delta} = \mathbb{C}$ and A(V) is a point.

<u>Example 9.10</u>. Let $f : V \longrightarrow W$ be a fibre space of complex manifolds. Suppose that the morphism f is of maximal rank at any point of V. Examples of such fibre spaces which are <u>not</u> fibre bundles are found in Atiyah [1], Kas [1], Kodaira [7] and Kuga [1]. Fix a point $w \in W$. The fundamental group $\pi_1(W, w)$ operates on $H^p(V_w, \mathbb{Q})$ where $V_w = f^{-1}(w)$. In addition, we have an isomorphism

$$H^0(W, R^p f_* \mathbb{Q}) \xrightarrow{\sim} H^p(V_w, \mathbb{Q})^{\pi_1(W, w)} .$$

On the other hand, there is a spectral sequence

$$H^p(W, R^q f_* \mathbb{Q}) \Longrightarrow H^{p+q}(V, \mathbb{Q}) .$$

Using the edge homomorphism of this spectral sequence, we obtain an exact sequence

(9.10.1) $\quad 0 \longrightarrow H^1(W, \mathbb{Q}) \longrightarrow H^1(V, \mathbb{Q}) \longrightarrow H^0(W, R^1 f_* \mathbb{Q})$.

Hence we have

$$b_1(V) \leqq b_1(W) + \dim_{\mathbb{Q}} H^1(V_w, \mathbb{Q})^{\pi_1(W, w)} .$$

Suppose, moreover, that V and W are <u>algebraic</u>. Deligne [2] has shown that, in this case, the third arrow of the above exact sequence (9.10.1) is surjective and we have

$$b_1(V) = b_1(W) + \dim_{\mathbb{Q}} H^1(V_w, \mathbb{Q})^{\pi_1(W, w)} .$$

By Corollary 9.3, we can conclude that we have

$$h^{1,0}(V) = h^{1,0}(W) + \dim_{\mathbb{C}} H^0(V_w, \Omega^1_{V_w})^{\pi_1(W,w)} .$$

Hence, if $H^0(V_w, \Omega^1_{V_w})^{\pi_1(W,w)} = 0$, from the proof of Theorem 9.7, we infer that $A(V)$ and $A(W)$ are isomorphic and that $V \xrightarrow{f} W \xrightarrow{\alpha} A(W)$ is nothing other than the Albanese mapping of V.

Lemma 9.11. Let V be a compact complex manifold and let $g : V \longrightarrow T$ be a meromorphic mapping of V into a complex torus. Then g is a morphism of V into T.

Proof. Let $\pi : \mathbb{C}^n \longrightarrow T$ be a universal covering of T. Let S be the set of points of indeterminacy of the meromorphic mapping g. For a point $p \in S$, we choose a simply-connected small open neighbourhood U of p in V. As S is of at least codimension two (Theorem 2.5), $U' = U - S \cap U$ is also simply-connected. Hence there exists a morphism $\tilde{g}' : U' \longrightarrow \mathbb{C}^n$ such that $\pi \cdot \tilde{g}' = g_{|U'}$. $\tilde{g}'$ can be considered as an n-tuple $(\tilde{g}_1', \tilde{g}_2', \cdots, \tilde{g}_n')$ of holomorphic functions on U'. As $S \cap U$ is of at least codimension two, by Hartogs' theorem, $\tilde{g}_i'$ can be extended to a holomorphic function $\tilde{g}_i$ on U. As $(\tilde{g}_1, \tilde{g}_2, \cdots, \tilde{g}_n)$ define a holomorphic mapping $\tilde{g} : U \longrightarrow \mathbb{C}^n$, $\pi \circ \tilde{g}$ defines a holomorphic mapping of U into T. Hence g can be naturally extended to a holomorphic mapping of V into T. Q.E.D.

Proposition 9.12. Let $f : \hat{V} \longrightarrow V$ be a bimeromorphic mapping of a complex manifold $\hat{V}$ onto a complex manifold V. We let

$(A(V), \alpha)$ be the Albanese torus of V. Then $(A(V), \alpha\circ f)$ is the Albanese torus of $\hat{V}$.

<u>Proof</u>. Let $(A(\hat{V}), \hat{\alpha})$ be the Albanese torus of $\hat{V}$. We choose a point $x \in \hat{V}$ and normalize the morphisms α and $\hat{\alpha}$ in such a way that $\hat{\alpha}(x)$ and $\alpha(y)$, $y = f(x)$, are origins of $A(\hat{V})$ and $A(V)$, respectively. Then, by Lemma 9.11, the meromorphic mapping $\alpha\circ f : \hat{V} \longrightarrow A(V)$ is a morphism. Using the universal property of the Albanese mapping, we have a group homomorphism

$$\begin{array}{ccc} \hat{V} & \xrightarrow{\hat{\alpha}} & A(\hat{V}) \\ f \downarrow & & \downarrow A(f) \\ V & \xrightarrow{\alpha} & A(V) \\ f^{-1} \downarrow & & \downarrow A(f^{-1}) \\ \hat{V} & \xrightarrow[\hat{\alpha}]{} & A(\hat{V}) \end{array}$$

$$A(f) : A(\hat{V}) \longrightarrow A(V) .$$

On the other hand, the meromorphic mapping $\hat{\alpha}\circ f^{-1} : V \longrightarrow A(\hat{V})$ is also a morphism because of Lemma 9.11. Hence there exists a group homomorphism

$$A(f^{-1}) : A(V) \longrightarrow A(\hat{V}) .$$

As we have $f\circ f^{-1} = id$ and $f^{-1}\circ f = id$, we obtain $A(f)\circ A(f^{-1}) = id$ and $A(f^{-1})\circ A(f) = id$. This implies that $A(f)$ is an isomorphism. Hence $(A(V), \alpha\circ f)$ is the Albanese torus of $\hat{V}$. Q.E.D.

<u>Definition 9.13</u>. Let $f : V \longrightarrow A$ be a morphism of a complex variety V into a complex torus A. We say (V, f) <u>generates</u> the complex torus A if there exists an integer n such that the morphism

$$\begin{array}{rcl} F : V\times V\times\cdots\times V & \longrightarrow & A \\ \psi & & \psi \\ (z_1, z_2, \cdots, z_n) & \longmapsto & f(z_1)+f(z_2)+\cdots+f(z_n) \end{array}$$

is surjective. We say that a subvariety B of a complex torus A generates A if (B, ι) generates A, where $\iota : B \longrightarrow A$ is the natural imbedding.

<u>Lemma 9.14</u>. For a complex manifold V, (V, α) generates the

Albanese torus $A(V)$ of V.

<u>Proof</u>. We define a morphism $\alpha^n : \underbrace{V \times \cdots \times V}_{n} \longrightarrow A(V)$ by

$$\alpha^n(z_1, z_2, \cdots, z_n) = \alpha(z_1) + \alpha(z_2) + \cdots + \alpha(z_n) .$$

If $\alpha_0 : V \longrightarrow A(V)$ is another Albanese mapping, we have

$$\alpha(z) = \alpha_0(z) + a ,$$

for a suitable $a \in A(V)$. Hence $\mathrm{Im}(\alpha^n) = \mathrm{Im}(\alpha_0^n) + na$. This implies that α^n is surjective if and only if α_0^n is surjective. Hence we can assume that there exists a point $z \in V$ such that $\alpha(z) = 0$. We set $A_n = \mathrm{Im}(\alpha^n)$. By Corollary 1.6, A_n is an irreducible subvariety of $A(V)$. As $0 \in A_1$, there is a sequence of inclusions

$$A_1 \subset A_2 \subset \cdots \subset A_n \subset A_{n+1} \subset \cdots .$$

Since each A_n is irreducible subvariety of $A(V)$, there exists a positive integer n_0 such that

$$A_{n_0} = A_{n_0+1} = \cdots = A .$$

By construction, for any $x, y \in A$, we have $x + y \in A$. Moreover, $0 \in A$. Now we shall prove that A is a complex subtorus of $A(V)$.

Let $\pi : \mathbb{C}^q \longrightarrow A(V)$ be a universal covering of $A(V)$ and let $\tilde{A}$ be the irreducible component of $\pi^{-1}(A)$ that contains the origin of $\mathbb{C}^q$. As A is irreducible, $\pi(\tilde{A}) = A$. For a point $\tilde{x} \in \tilde{A}$, let $\ell_{\tilde{x}}$ be a complex line which goes through $\tilde{x}$ and the origin. We shall prove that $\ell_{\tilde{x}} \subset \tilde{A}$. For that purpose we first remark that for any positive integer m, the multiplication of A by m maps A <u>onto</u> A. This implies that, for any $x \in A$, there exists $y \in A$ such that $my = x$. Therefore, for sufficiently large positive integers n, $\frac{1}{n}\tilde{x} \in \tilde{A}$. Let $I(\tilde{A})$ be the defining ideal of $\tilde{A}$ in $\mathbb{C}^q$.

For any holomorphic function in $I(\widetilde{A})$, we have $f(\frac{1}{n}\tilde{x}) = 0$ for sufficiently large n. It follows that f vanishes in a neighbourhood of the origin of $\ell_{\tilde{x}}$ and, thus, that f vanishes on ℓ_{x}. Hence $\ell_{\tilde{x}} \subset \widetilde{A}$. This implies that for $\tilde{x} \in \widetilde{A}$, $-\tilde{x} \in \hat{A}$. Thus, for any $x \in A$, $-x \in A$. We conclude that A is a group subvariety of $A(V)$ and, therefore, a complex subtorus of $A(V)$.

Let $g : V \longrightarrow B$ be a morphism of V into a complex torus B such that $g(z) = 0$, where $z \in V$ and $\alpha(z) = 0$. Then there exists a group homomorphims $h : A(V) \longrightarrow B$ such that $h \circ \alpha = g$. We set

$$\begin{array}{ccc} V & \xrightarrow{\alpha} & A \subset A(V) \\ & {\scriptstyle g}\searrow \quad \swarrow{\scriptstyle h} & \\ & B & \end{array}$$

$h_A = h|_A : A \longrightarrow B$. h_A is also a group homomorphism. As $\alpha(V) \subset A$, we have

$$h_A \circ \alpha = g .$$

Hence (A, α) satisfies the universal properties of the Albanese torus. This implies that $A = A(V)$. Q.E.D.

Using Lemma 9.14 we prove the following :

<u>Proposition 9.15</u>. If V is a Moishezon manifold, $A(V)$ is an abelian variety.

<u>Proof</u>. By Lemma 9.14, there exists a positive integer n such that

$$\begin{array}{rcl} \alpha^n : V \times V \times \cdots \times V & \longrightarrow & A(V) \\ \Psi & & \Psi \\ (z_1, z_2, \cdots, z_n) & \longmapsto & \alpha(z_1)+\alpha(z_2)+\cdots+\alpha(z_n) \end{array}$$

is surjective. The assumption that V is Moishezon implies that $V \times \cdots \times V$ is also Moishezon. By Corollary 3.9, $A(V)$ is Moishezon. On the other hand, a complex torus A is an abelian variety if and only if A is Moishezon (see Siegel [1], Weil [2], Théorème 3 p.124).

Q.E.D.

Remark 9.16. From the bimeromorphic invariance of Albanese tori, the above proposition is reduced to the case where V is projective algebraic. Then, applying the theory of harmonic integrals, we can explicitly show that $A(V)$ is abelian if V is projective (see Weil [2], p.82).

In the following sections we shall show that the study of Albanese mappings are very important for the classification theory. Unfortunately, fibres of the Albanese mapping $\alpha : V \longrightarrow \alpha(V) \subset A(V)$ are not necessarily connected (see the following Example 9.18). Therefore we introduce the notion "the fibre space associated with the Albanese mapping".

Definition 9.17. For the Albanese mapping $\alpha : V \longrightarrow A(V)$, let

$$\begin{array}{ccc} V & \xrightarrow{\alpha} & \alpha(V) \\ & {\scriptstyle\beta}\searrow \quad \nearrow{\scriptstyle\gamma} & \\ & W & \end{array}$$

be the Stein factorization of $\alpha : V \longrightarrow \alpha(V)$. The fibre space $\beta : V \longrightarrow W$ is called the fibre space associated with the Albanese mapping.

Note that, in view of Proposition 6.10,1), $\kappa(W) \geqq \kappa(\alpha(V))$.

Example 9.18. Let A be a two-dimensional abelian variety and let D be an ample divisor on A. We set $V = \mathbb{P}^1 \times A$. Then, for a point $p \in \mathbb{P}^1$, a divisor $E = p \times A + \mathbb{P}^1 \times D$ is ample. For a sufficiently large n, $\Phi_{nE} : V \longrightarrow \mathbb{P}^N$ gives a projective imbedding. Let S be a non-singular surface cut out on $\Phi_{nE}(V)$ by a general hyperplane of $\mathbb{P}^N$. By the Lefschetz theorem on hyperplane sections, $H_1(S, \mathbb{Z}) \simeq H_1(V, \mathbb{Z})$ and $H^0(S, \Omega^1_S)$ is induced from $H^0(V, \Omega^1_V)$.

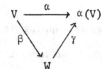

Moreover, $H^0(V, \Omega_V^1) \simeq H^0(A, \Omega_V^1)$ and $H_1(V, \mathbb{Z}) \rightrightarrows H_1(A, \mathbb{Z})$. This implies that $A(S) \simeq A(V) = A$. Hence the Albanese mapping $\alpha : S \longrightarrow A$ is surjective. If we know that $\kappa(S) > 0$, then, as $\kappa(A) = 0$, α is not an isomorphism. In that case, the general fibre of α is not connected. Now we shall show that $\kappa(S) = 2$. By the adjunction formula (Lemma 6.8), $K_S = K_V|_S \otimes [S]|_S$. On the other hand $K_V = [-2p \times A]$ and $[S] = [nE]$. Hence $K_S = [(n-2)p \times A + n\mathbb{P}^1 \times D]|_S$. As K_S is ample, we obtain $\kappa(S) = 2$.

Let M be a simply connected compact complex manifold. Then $A(S \times M) = A(S)$. Hence a general fibre of the Albanese mapping $\alpha : S \times M \longrightarrow A(V) = A$ is not connected. The fibre space associated with the Albanese mapping α is nothing other than the projection $p : S \times M \longrightarrow S$ of $S \times M$ onto the first factor.

If the image of the Albanese mapping is a curve, fibres of the Albanese mapping are connected. We have the following :

Proposition 9.19. Assume that the image of the Albanese mapping $\alpha : V \longrightarrow A(V)$ is a curve C. Then C is a non-singular curve of genus $g = \dim A(V)$ and the fibres of $\alpha : V \longrightarrow C$ are connected.

Proof. Let $f : \tilde{C} \longrightarrow C$ be a normalization of the curve C and let $J(\tilde{C})$ be the Jacobian variety of $\tilde{C}$. We let $g : \tilde{C} \longrightarrow J(\tilde{C})$ be the canonical morphism. A meromorphic mapping $\tilde{\alpha} = g \circ f^{-1} \circ \alpha : V \longrightarrow J(\tilde{C})$ is holomorphic by virtue of Lemma 9.11. By using the universal property of $(A(V), \alpha)$, it is easy to show that $(J(\tilde{C}), \tilde{\alpha})$ satisfies the universal property of the Albanese torus of V. Hence f must be an isomorphism and C is a non-singular curve of a genus $g = \dim A(V)$.

Suppose that a fibre of α is not connected. Let

$$\begin{array}{ccc} V & \xrightarrow{\alpha} & C \\ & {\scriptstyle\beta}\searrow \quad \nearrow{\scriptstyle\gamma} & \\ & W & \end{array}$$

be the Stein factorization of α. $\gamma : W \longrightarrow C$ is a finite morphism. As C is non-singular, W is non-singular. First we consider the case where the genus $g(W)$ of W is strictly larger than the genus $g(C)$ of C. Let $h : W \longrightarrow J(W)$ be the canonical imbedding of W into the Jacobian of W. Then, by the universal mapping property of the Albanese torus, there exists a group homomorphism $j : A(C) \longrightarrow J(W)$ such that $h \circ \beta = j \circ \alpha$. Hence there exists a surjective morphism

$$\begin{array}{ccc} V & \xrightarrow{\alpha} & C \hookrightarrow A(C) = J(C) \\ {\scriptstyle\beta}\searrow & & \downarrow j \\ W & \xrightarrow{h} & J(W) \end{array}$$

$j_{|C} : C \longrightarrow W$. This contradicts the fact that $g(W) > g(C)$.

Next, if $g(C) = g(W)$, then C and W are elliptic curves and γ is unramified. Then it is easy to show that $(J(W),\ h \circ \beta)$ satisfies the universal property of the Albanese torus of V. Hence γ is an isomorphism. Q.E.D.

Finally we introduce certain bimeromorphic invariants of a complex variety, which play an important role in our classification theory.

Definition 9.20. Let V be a complex variety of dimension n and let V^* be a non-singular model of the variety V. We define

$$g_k(V) = \dim_{\mathbb{C}} H^0(V^*, \Omega^k_{V^*}), \quad k = 1, 2, \cdots, n ,$$

$$r(V) = \dim_{\mathbb{C}} H^0(V^*, d\underline{O}_{V^*}),$$

$$q_k(V) = \dim_{\mathbb{C}} H^k(V^*, \underline{O}_{V^*}), \quad k = 1, 2, \cdots, n ,$$

$$q(V) = q_1(V) .$$

These are independent of the choice of a non-singular model V^* of V and are bimeromorphic invariants of a complex variety. (For $g_k(V)$ and $r(V)$, the proof is similar to that of Lemma 6.3. For $q_k(V)$, see Corollary 2.15.) $q_k(V)$ is called the <u>k-th irregularity</u> of the variety V. $q(V)$ is usually called the <u>irregularity</u> of the variety V. By the Serre duality, we have

$$q_n(V) = g_n(V) = p_g(V) .$$

<u>Defition 9.21</u>. The <u>Albanese dimension</u> $t(V)$ of a complex variety V is the dimension of the Albanese torus $A(V^*)$ of a non-singular model V^* of the variety V.

By virtue of Proposition 9.12, this is well defined and the Albanese dimension $t(V)$ is also a bimeromorphic invariant of the variety V. Corollary 9.3 and Corollary 9.7, 1) imply the following :

<u>Lemma 9.22</u>. For a complex variety V, we have

$$t(V) \leqq r(V) \leqq g_1(V).$$

Moreover, if V has a non-singular model V^* which satisfies the assumption of Corollary 9.3 (for example, V is a Moishezon variety or V is a Kähler manifold), then, a fortiori, in the above inequalities, the equalities hold and

$$q_k(V) = g_k(V), \quad k = 1, 2, 3, \cdots , n .$$

The following lemma is an easy consequence of Definition 9.20 and Definition 9.21.

<u>Lemma 9.23</u>. For a surjective morphism $f : V \longrightarrow W$ of complex varieties, we have inequalities

$$g_k(V) \geqq g_k(W) , \quad k = 1, 2, \cdots , \dim W,$$

$$r(V) \geqq r(W) ,$$

$$t(V) \geqq t(W) .$$

The following proposition is due to Freitag.

<u>Proposition 9.24</u>. Let V be a complex manifold of dimension n and let G be a finite group of analytic automorphisms of V. Then we have

$$g_k(V/G) = \dim_{\mathbb{C}} H^0(V, \Omega_V^k)^G, \quad k = 1, 2, \cdots, n .$$

For the proof, see Freitag [1], Satz 1, p.99 .

<u>Remark 9.25</u>. In the same situation as in Proposition 9.24, it is easy to show that

$$P_m(V/G) \leqq \dim_{\mathbb{C}} H^0(V, \underline{O}(mK_V))^G .$$

If $m \geqq 2$, the equality does <u>not</u> necessarily hold.

§ 10. Subvarieties of complex tori

To study the fibre space $\beta : V \longrightarrow W$ associated with the Albanese mapping $\alpha : V \longrightarrow A(V)$, it is natural to study the image variety $\alpha(V)$ of α. More generally, in this section we study structures of subvarieties of complex tori.

Let A be a complex torus of dimension n. By global coordinates of A, we shall mean global coordinates of the universal covering $\mathbb{C}^n$ of A such that, in terms of these coordinates, covering transformations are represented by translations by elements of a lattice Δ in $\mathbb{C}^n$, where $A = \mathbb{C}^n/\Delta$.

Lemma 10.1. Let B be an ℓ-dimensional subvariety of a complex torus A. Then we have

$$g_k(B) \geqq \binom{\ell}{k}, \qquad k = 1, 2, \cdots, \ell,$$

$$P_m(B) \geqq 1, \qquad m = 1, 2, \cdots .$$

Hence, a fortiori, we have

$$\kappa(B) \geqq 0, \quad g_1(B) \geqq \dim B .$$

Proof. Let $\tau : B^* \longrightarrow B$ be a desingularization of the variety B. There exists a point $p \in B$ such that B is smooth at p and

$$z_1 - z_1^*, \ z_2 - z_2^*, \ \cdots, \ z_\ell - z_\ell^*$$

induce local coordinates in B with center p where $(z_1, z_2, \cdots, z_\ell, z_{\ell+1}, \cdots, z_n)$ is a system of global coordinates of A and $(z_1^*, z_2^*, \cdots, z_n^*)$ is a point of $\mathbb{C}^n$ lying over the point p.

In this situation,

$$\tau^*(dz_{i_1} \wedge \cdots \wedge dz_{i_k}), \quad 1 \leqq i_1 < \cdots < i_k \leqq \ell,$$

is a non-zero holomorphic k-form on B^*. Hence $g_k(B) \geqq \binom{\ell}{k}$. Moreover, $\tau^*((dz_1 \wedge \cdots \wedge dz_\ell)^m)$ is an element of $H^0(B^*, \underline{O}(mK_{B^*}))$.

Q.E.D.

Corollary 10.2. Let B be a subvariety of a complex torus. Then B is neither unirational nor ruled.

Now we shall characterize a subvariety of a complex torus of Kodaira dimension zero.

Theorem 10.3. Let B be an ℓ-dimensional subvariety of a complex torus A. Then the following conditions are equivalent.

1) $g_k(B) = \binom{\ell}{k}$ for a positive integer k, $1 \leqq k \leqq \ell$.

2) $p_g(B) = 1$.

3) $P_m(B) = 1$ for a positive integer m.

4) $\kappa(B) = 0$

5) B is a translation of a complex subtorus A_1 of A by an element $a \in A$.

Proof. It is clear that 5) implies 1), 2), 3) and 4), also 4) implies 3). In general, for a complex manifold V, if $p_g(V) \geqq 2$ then $P_m(V) \geqq 2$, $m \geqq 2$, because $(\varphi_1)^m$, $(\varphi_2)^m \in H^0(V, \underline{O}(m K_V))$ are linearly independent if φ_1, φ_2 are linearly independent elements of $H^0(V, \underline{O}(K_V))$. Hence, 3) implies 2). Moreover, 2) is a special case of 1). Thus, it suffices to show that 1) implies 5).

We can assume that B contains the origin 0 of a complex torus A and that B is smooth at 0. In addition, we can choose a system of global coordinates $(z_1, z_2, \ldots, z_n)$ of A such that $(z_1, z_2, \ldots, z_\ell)$ gives a system of local coordinates in B with center 0.

Hence, in a small neighbourhood U of the origin 0 in A, the variety B is defined by the equations

(10.3.1) $z_j = f_j(z_1, z_2, \cdots, z_\ell), \quad j = \ell+1, \ell+2, \ldots, n,$

where the f_j's are holomorphic functions on $U \cap B$.

We shall show that f_k's are linear functions of $z_1, z_2, \cdots, z_\ell$.

Let $\tau : B^* \longrightarrow B$ be a desingularization of B. As $g_k(B) = \binom{\ell}{k}$ for an k, $1 \leq k \leq \ell$, $\{\tau^*(dz_{i_1} \wedge \cdots \wedge dz_{i_k}), 1 \leq i_1 \ldots i_k \leq \ell\}$ is a basis of $H^0(B^*, \Omega^k_{B^*})$. It follows that

$$\tau^*(dz_{j_1} \wedge \cdots \wedge dz_{j_k}) = \sum_{(i)} C_{(i)(j)} \tau^*(dz_{i_1} \wedge \cdots \wedge dz_{i_k}),$$

where $(j) = (j_1, \cdots, j_k)$, $1 \leq j_1 < \cdots < j_k \leq n$, $(i) = (i_1, \cdots, i_k)$, $1 \leq i_1 < \cdots < i_k \leq \ell$ and $C_{(i)(j)}$ is a constant.

Taking an exterior product by $\tau^*(dz_{k+1} \wedge \cdots \wedge dz_\ell)$ on both side of the last equality, we obtain

$$\tau^*(dz_1 \wedge \cdots \wedge \check{dz_i} \wedge \cdots \wedge dz_\ell \wedge dz_j)$$

$$= C_{ij} \tau^*(dz_1 \wedge \cdots \wedge dz_\ell),$$

$$i = 1, 2, \cdots, \ell, \quad j = \ell+1, \ell+2, \cdots, n,$$

where C_{ij} is a constant.

On the other hand, from (10.3.1), we derive

$$\tau^*(dz_1 \wedge \cdots \wedge \check{dz_i} \wedge \cdots \wedge dz_\ell \wedge dz_j) = (-1)^{\ell-i} \frac{\partial f_j}{\partial z_i} \tau^*(dz_1 \wedge \cdots \wedge dz_\ell),$$

on $\tau^{-1}(U \cap B)$. This implies that $\dfrac{\partial f_j}{\partial z_i} = (-1)^{\ell-i} C_{ij}$.

Therefore, the f_j's are linear functions.

Let L be a linear subspace of $\mathbb{C}^n$ defined by equations (10.3.1).

Pick the irreducible components $\tilde{B}$ of the inverse image of B in the universal covering $\mathbb{C}^n$ which coincides with L in a small neighbourhood of 0 in $\mathbb{C}^n$. As $\tilde{B}$ and L are irreducible, $\tilde{B} = L$. This implies that B is a complex subtorus. Q.E.D.

From the above proof we conclude the following :

Corollary 10.4. Let B be the same as above and let $\tau : B^* \longrightarrow B$ be a desingularization of B. Assume that

$$\dim_{\mathbb{C}} \tau^*(H^0(A, \Omega^k_A)) = \binom{\ell}{k} .$$

Then B is non-singular and is a translation of a complex subtorus of A by an element of A.

Corollary 10.5. If a proper subvariety B of a complex torus A generates A, then we have

$$\kappa(B) > 0 .$$

Proof. Assume the contrary. Then B is a translation of a complex subtorus. Hence it is impossible that B generates A. Q.E.D.

As the consequence of Corollary 10.5 and Lemma 9.14, the following results hold.

Corollary 10.6. Let V be a complex manifold and let $\alpha : V \longrightarrow A(V)$ be the Albanese mapping of V. Then we have

$$\kappa(\alpha(V)) \geqq 0 .$$

Moreover, $\kappa(\alpha(V)) = 0$ if and only if the Albanese mapping α is surjective.

Corollary 10.7. Let B be a subvariety of a complex torus.

If $g_1(B) > \dim B$, then

$$\kappa(B) > 0.$$

Next we shall study the structures of subvarieties of positive Kodaira dimensions of a complex torus. First we shall prove the following lemma.

Lemma 10.8. Let B be a subvariety of complex torus A such that $a(B) = \dim B$. Then B is a subvariety of an abelian variety A_1 which is a complex subtorus of A.

Proof. We define a morphism

$$\begin{array}{cccc} f_n : & \overbrace{B \times B \times \cdots \times B}^{2n} & \longrightarrow & A \\ & \text{\rotatebox{90}{$\in$}} & & \text{\rotatebox{90}{$\in$}} \\ & (z_1, z_2, \ldots, z_{2n}) & \longmapsto & z_1+z_2+\cdots+z_n-(z_{n+1}+\cdots+z_{2n}) \end{array}$$

for each positive integer n. Let F_n be the image of f_n. F_n is a subvariety of A and we have a sequence of inclusions

$$F_1 \subset F_2 \subset \cdots \subset F_n \subset F_{n+1} \subset \cdots .$$

There exists a positive integer n_0 such that

$$F_{n_0} = F_{n_0+1} = \cdots = A_1 .$$

As A_1 is closed under the addition and the subtraction induced from those of A, A_1 is a complex subtorus. By Corollary 3.9,2) and Proposition 9.15, A_1 is an abelian variety. It is easy to show that A_1 is the desired one. Q.E.D.

Theorem 10.9. Let B be a subvariety of a complex torus A. Then there exist a complex subtorus A_1 of A and a projective variety W which is a subvariety of an abelian variety such that

1) B is an analytic fibre bundle over W whose fibre is A_1;

2) $\kappa(W) = \dim W = \kappa(B)$.

Furthermore, if B is an algebraic variety, then there exist finite unramified coverings $\tilde{B}$ and $\tilde{W}$ of B and W, respectively, such that

$$\tilde{B} = A_1 \times \tilde{W} .$$

Proof. We can assume that $\kappa(B) > 0$. By Theorem 6.11, there exist a desingularization $\tau : B^* \longrightarrow B$, a smooth projective variety W^* and a surjective morphism $f : B^* \longrightarrow W^*$, which satisfy the conditions 1) ~ 5) in Theorem 6.11. There exists a dense subset U^* of W^* such that $B^*_w = f^{-1}(w)$ is smooth and $\kappa(B^*_w) = 0$ for all $w \in U^*$. We set $B_w = \tau(B^*_w)$. Then there exists a dense subset $U^\#$ of U^* such that B_w is bimeromorphically equivalent to B^*_w for any point $w \in U^\#$. As we have $\kappa(B_w) = \kappa(B^*_w) = 0$ for $w \in U^\#$, B_w is a translation of a complex subtorus A_w of dimension $\ell = \dim B - \kappa(B)$, by Theorem 10.3. As any deformation of a complex torus is a complex torus (see Andreotti and Stoll [2], Theorem 8, p.339), any regular fibre of τ^* is bimeromorphically equivalent to a complex torus. There is a connected open dense set U of W^* such that, for any point $w \in U$, B^*_w is bimeromorphically equivalent to B_w. Since $f^{-1}(U) \longrightarrow U$ is a complex analytic family whose fibres are bimeromorphically equivalent to complex tori, all complex subtori A_w bimeromorphically equivalent to B^*_w, $w \in U$, are isomorphic to a complex subtorus A_1 of A. (Note that any complex torus contains at most countably many complex subtori.)

We set $A_2 = A/A_1$ and $u : A \longrightarrow A_2$ is the canonical quotient map. We set $g = u \circ \tau$.

Let X be the image of B^* in $W^* \times A_2$ of the morphism (f, g) : $B^* \longrightarrow W^* \times A_2$ and let $h : B^* \longrightarrow X$ be the canonical morphism.

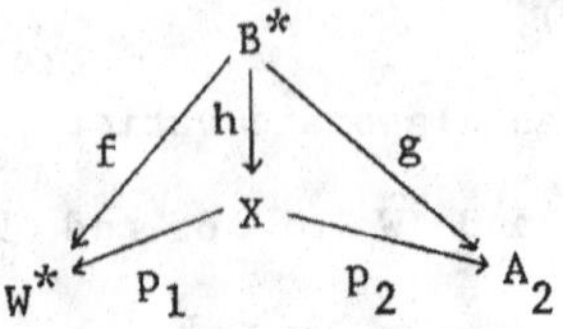

Let $p_1 : X \longrightarrow W^*$ and $p_2 : X \longrightarrow A_2$ be the morphisms induced from the projections to the first factor and the second factor of $W^* \times A_2$, respectively. Since, for any point $w \in U$, $g(f^{-1}(w))$ is a point, $p_1^{-1}(U)$ is isomorphic to U. Hence $\dim X = \dim W^*$. We set $W = p_2(X)$. W is a subvariety of a complex torus A_2. We have $\dim W \leqq \dim W^*$. We set $S = g(f^{-1}(U))$. $S \subset W$. As $(p_2 \circ h)^{-1}(S) = g^{-1}(S)$ is open dense in B^*, we have

$$\dim W + \dim A_1 \geqq \dim B^*.$$

It follows that $\dim W = \dim W^*$. Hence, $\dim u^{-1}(W) = \dim W + \dim A_1 = \dim B$. As $u^{-1}(W) \supset B$, we conclude that $u^{-1}(W) = B$. Thus, $u : B \longrightarrow W$ must be an analytic fibre bundle over W, whose fibre is a complex torus A_1.

Next we prove that $\kappa(W) = \dim W = \kappa(B)$.
If this is true, then $a(W) = \dim W$. Hence, by Lemma 10.8, W is contained in an abelian variety.

Now suppose that $\kappa(W) < \dim W$. The above arguments shows that there exist a subvariety W_1 of a complex torus and a surjective morphism $u_1 : W \longrightarrow W_1$ which is an analytic fibre bundle over W_1 whose fibre is a complex torus. Then $u_1 \circ u : B \longrightarrow W_1$ is an analytic fibre bundle whose fibre F is a torus bundle over a torus. In view of Example 6.9,2), $\kappa(F) \leqq 0$. Hence, by Theorem 6.12, we have

$$\kappa(B) \leqq \kappa(F) + \dim W_1 < \dim W.$$

This contradicts the fact that $\kappa(B) = \dim W$.

Finally, assume that the variety B is algebraic. Lemma 10.8, implies that B is a subvariety of an Abelian variety. Hence, we can assume that A is an abelian variety. In the above arguments $u : A \longrightarrow A_2$ becomes a fibre bundle in the etale topology, i.e. there exists finite unramified coverings $\pi : \tilde{A} \longrightarrow A$ and $\pi_2 : \tilde{A}_2 \longrightarrow A_2$ such that $\tilde{A} = A_1 \times \tilde{A}_2$ (Poincaré's reducibility theorem; see Mumford [9] Theorem 1, p.173). Let $\tilde{B}$ and $\tilde{W}$ be one of the irreducible component of $\pi^{-1}(B)$ and $\pi_2^{-1}(W)$, respectively, such that $\tilde{p}_2(\tilde{B}) = \tilde{W}$, where $\tilde{p}_2 : \tilde{A} \longrightarrow \tilde{A}_2$ is the natural projection to the second factor. Then $\tilde{B} = A_1 \times \tilde{W}$. Q.E.D.

Corollary 10.10. Let B be a proper subvariety of a simple abelian variety. Then $\kappa(B) = \dim B$.

Lemma 10.11. Let B be a smooth subvariety of a complex torus. Then the sheaf $\underline{O}(mK_B)$ is spanned by its global sections for any positive integer m. Thus, if $\kappa(B) > 0$, the complete linear system $|mK_B|$ is free from base points and fixed components.

Proof. We have an exact sequence

$$0 \longrightarrow T_B \longrightarrow T_{A|B} \longrightarrow N_B \longrightarrow 0 \quad ,$$

where T_A, T_B are tangent bundle of A and B, respectively, and N_B is the normal bundle of B in A. As $\underline{O}(T_A)$ is spanned by its global sections, $\underline{O}(N_B)$ is spanned by its global sections. Hence $\underline{O}(K_B) = \underline{O}(\overset{\ell}{\wedge} N_B)$, $\ell = \operatorname{codim} B$, is spanned by its global sections and, a fortiori, $\underline{O}(mK_B)$ is spanned by its global sections. Q.E.D.

Corollary 10.12. If B is a smooth subvariety of a complex torus, the structure of the fibre bundle in Theorem 10.9 is bimeromorphically equivalent to the fibre space given by the m-th canonical map $\Phi_{mK} : B \longrightarrow W = \Phi_{mK}(B)$ for a sufficiently large m.

Proof. By virtue of Lemma 10.11, in the proof of Theorem 10.9 we can use the fibre space $\Phi_{mK} : B \longrightarrow W$. Then the arguments in the proof of Theorem 10.9 imply the desired result. Q.E.D.

Remark 10.13. Hartshorne [2] has shown that the canonical bundle K_W of a submanifold W of a simple abelian variety is ample. Ueno [2] claimed that if $\kappa(W) = \dim W$ for a submanifold of a complex torus, then the canonical bundle K_W is ample. Unfortunately there is a gap in the proof and the problem whether K_W is ample or not is still open.

§ 11. Classification theory

In this section, we shall give an outline of our classification theory and discuss related conjectures and problems.

Definition 11.1. A complex variety V is called a variety of hyperbolic type (parabolic type, elliptic type, respectively) if $\kappa(V) = \dim V$ ($\kappa(V) = 0$, $\kappa(V) = -\infty$, respectively).

The fundamental theorem on the pluricanonical fibrations (Theorem 6.11) says that the classification theory is reduced to

1) the study of complex varieties of hyperbolic, parabolic and elliptic type ;

2) the study of fibre spaces whose general fibres are of parabolic type.

We shall show below that the study of complex varieties of parabolic and elliptic type is deeply related to the study of fibre spaces associated with Albanese mappings (Definition 9.17).

(11.2). First we shall consider a complex manifold V of hyperbolic type. Since $a(V) \geqq \kappa(V) = \dim V$, V is a Moishezon manifold. By Theorem 6.11, the pluricanonical mapping Φ_{mK} is bimeromorphic for a sufficiently large $m \in \mathbb{N}(V)$. If V is a curve, then V is of hyperbolic type if and only if the genus of the curve V is greater than or equal to two. In this case, the canonical bundle K_V is ample and $3K_V$ is very ample. Moreover, if V is not a hyperelliptic curve, K_V is very ample. A surface S of hyperbolic type is usually called a surface of general type. If S is a minimal surface of general type (see Definition 20.2 and Theorem

20.3, below), then the pluricanonical system $|mK|$ is free from base points and fixed components for $m \geqq 4$ and Φ_{mK} is a birational morphism for $m \geqq 5$ (see Kodaira [4] and Bombieri [1].) It is natural to ask whether these properties hold for a manifold of hyperbolic type of dimension $n \geqq 3$. There exists a three-dimensional algebraic manifold V of hyperbolic type such that the pluricanonical system $|mK(V^*)|$ of any birationally equivalent model V^* of V has always fixed components for any $m \geqq 1$ (see Ueno [7] and Example 16.18 below).

The following problems on manifolds of hyperbolic type are important.

<u>Problem 11.2.1</u>. Does there exist a bimeromorphically equivalent model V^* of a complex manifold V of hyperbolic type such that the pluricanonical system $|mK(V^*)|$ is free from base points (but may have fixed components) for a sufficiently large integer m ?

<u>Problem 11.2.2</u>. Does there exist a positive integer m_0 which only depends on the dimension of V such that, for any integer $m \geqq m_0$, the m-th canonical mapping Φ_{mK} is bimeromorphic ?

<u>Problem 11.2.3</u>. Is any deformation of a complex manifold of hyperbolic type again a complex manifold of hyperbolic type ?

Any deformation of a surface of general type is again a surface of general type (see Kodaira [3], I, Theorem 24 and Iitaka [1], II).

<u>Problem 11.2.4</u>. Does there exist a moduli space (a coarse moduli space in the sense of Mumford [5]) of complex manifolds of hyperbolic type ? If the moduli space exists, is it an algebraic space ?

Popp [1], II has shown that the moduli space of surfaces of general type exists as an algebraic space (see Theorem 13.10, below). Moreover, if the canonical bundle K_V of V is ample, Matsusaka [1] has shown that there exists a positive integer m_0 which depends on the Hilbert polynomial $p(m) = \sum_{k=0}^{n} (-1)^k \dim H^k(V, \underline{O}(mK))$ such that for any integer $m \geqq m_0$, mK_V is very ample. This is a partial answer to Problem 11.2.1 and Problem 11.2.2. Using this fact and Popp [1] II we can show the following :

Theorem 11.2.5. Canonically polarized algebraic manifolds have moduli spaces as algebraic spaces.

The existence of moduli spaces of canonically polarized algebraic manifolds as complex spaces has been proved by Narashimhan and Simha [1].

Problem 11.2.4 is deeply related to the problem of classification of complex manifolds of algebraic dimension zero (see §13, especially the proof of Theorem 13.11).

In §14 below, we shall show that the group Bim(V) of bimeromorphic mappings of a complex variety V of hyperbolic type onto itself is finite (see Theorem 14.3).

(11.3) Next we shall consider a complex manifold of parabolic type. If a certain multiple of the canonical bundle K_V of a manifold V is trivial, then V is of parabolic type. In §17, we shall show that there are pathological examples of no-Kähler manifolds with trivial canonical bundles. Hence, here, we restrict our attention to algebraic manifolds of parabolic type.

The most important result in this direction is the following theorem due to Matsushima.

Theorem 11.3.1. Let V be a projective manifold with $c_1(V) = 0$ ($c_1(V)$ is the first Chern class of V considered as an element in $H^2(V, \mathbb{Q})$). Then there exist an abelian variety A and a projective manifold F such that

1) $c_1(F) = 0$ and $q(F) = 0$;

2) $A \times F$ is a finite unramified Galois covering of V with a finite solvable Galois group.

For the proof, see Matsushima [2],[3]. Calabi [1] has shown that the above theorem holds for any Kähler manifold if his famous conjecture is true.

Corollary 11.3.2. Let V be a projective manifold with $c_1(V) = 0$ in $H^2(V, \mathbb{Q})$. There exists a positive integer m such that mK_V is analytically trivial.

Proof. Let $\pi : \tilde{V} \longrightarrow V$ be an n-fold unramified covering of V. First we shall show that, if $\ell K_{\tilde{V}}$ is trivial for a positive integer ℓ, then $\ell n K_V$ is trivial. As we have $K_{\tilde{V}} = f^* K_V$, we have $f_*\underline{O}(mK_{\tilde{V}}) = \underline{O}(mK_V) \otimes f_*\underline{O}_V$. By our assumption, we have $\underline{O}(\ell K_{\tilde{V}}) = \underline{O}_{\tilde{V}}$. Hence we obtain

$$\overset{n}{\wedge} f_*\underline{O}_{\tilde{V}} = \overset{n}{\wedge} f_*\underline{O}(\ell K_{\tilde{V}}) = \overset{n}{\wedge}(\underline{O}(\ell K_V) \otimes f_*\underline{O}_{\tilde{V}})$$
$$= \underline{O}(\ell n K_V) \otimes \overset{n}{\wedge} f_*\underline{O}_{\tilde{V}} .$$

Since $f_*\underline{O}_{\tilde{V}}$ is a locally free sheaf of rank n, $\underline{O}(\ell n K_V)$ is isomorphic to $\underline{O}_V$.

Next we shall show that if $c_1(V) = 0$ in $H^2(V, \mathbb{Q})$ and $q(V) = 0$, then there exists a positive integer m such that mK_V is trivial. From an exact sequence

$$0 \longrightarrow \mathbf{Z} \longrightarrow \underline{O}_V \xrightarrow{\exp} \underline{O}_V^* \longrightarrow 0,$$

We have a long exact sequence

$$\longrightarrow H^1(V, \underline{O}_V) \longrightarrow H^1(V, \underline{O}_V^*) \xrightarrow{c} H^2(V, \mathbf{Z}) \longrightarrow .$$

Let η be the element of $H^1(V, \underline{O}_V^*)$ corresponding to the canonical bundle K_V of V. Then $c_1(V) = -c(\eta)$ in $H^2(V, \mathbf{Z})$. Since $c_1(V) = 0$ in $H^2(V, \mathbf{Q})$, there exists a positive integer m such that $mc(\eta) = c(\eta^m) = 0$ in $H^2(V, \mathbf{Z})$. As $H^1(V, \underline{O}_V) = 0$, c is injective. Hence η^m is zero in $H^1(V, \underline{O}_V^*)$. That is, mK_V is trivial.

Let V be a projective manifold with $c_1(V) = 0$ in $H^2(V, \mathbf{Q})$. By Theorem 11.3.1, there exist a projective manifold F and an abelian variety A which satisfy the conditions 1), 2) in Theorem 11.3.1. Then from the above arguments, we infer that there exists a positive integer m such that mK_V is trivial. Q.E.D.

In §16, we shall show that there exist many projective manifolds V of parabolic type such that for any integer m, mK_{V^*} is <u>not</u> trivial for any birationally equivalent model V^* of V (see Corollary 16.11.3, Corollary 16.12.3 and Proposition 16.17). Hence, contrary to the case of surfaces (see §20 ; for a minimal surface S of parabolic type, $12K_S$ is analytically trivial), Theorem 11.3.1 is not sufficient to study the structure of algebraic manifolds of parabolic type.

(11.4). There are a lot of conjectures concerning algebraic manifolds of parabolic type. Here we shall give the most important conjectures.

<u>Conjecture Q_n</u>. If V is an n-dimensional algebraic manifold of parabolic type, then $q(V) \leq n$.

This is an easy consequence of the following conjecture.

Conjecture A_n. Suppose that V is the same as in Conjecture Q_n. The Albanese mapping $\alpha : V \longrightarrow A(V)$ is surjective.

Iitaka [3] has introduced the following birational invariant of an algebraic variety of parabolic type.

Definition 11.4.1. The maximal irregularity $q^*(V)$ of an algebraic variety V of parabolic type is defined by

$$q^*(V) = \operatorname*{Sup}_{\tilde{V}} q(\tilde{V}),$$

where $\tilde{V}$ runs through the set of all finite unramified coverings of V.

If Conjecture Q_n is true, the maximal irregularity $q^*(V)$ is finite. It is possible that we have $q^*(V) > 0$ even if $q(V) = 0$ (see Example 16.19). Using the notion of the maximal irregularity, we have the following :

Conjecture Q_n^*. For an n-dimensional algebraic manifold V of parabolic type, we have $q^*(V) \neq n - 1$.

Conjecture Q_n and Conjecture Q_n^* are found in Iitaka [3]. A more geometric conjecture corresponding to Conjecture Q_n^* is the following.

Conjecture A_n^*. Let V be an n-dimensional algebraic manifold of parabolic type with $q(V) = n - 1$. There exists a finite unramified covering $\tilde{V}$ of V such that $\tilde{V}$ is birationally equivalent to a product of an (n-1)-dimensional abelian variety and an elliptic curve.

Ueno [3] has proposed the following conjecture including all of the above conjectures.

Conjecture K_n. Let V be an n-dimensional algebraic manifold of parabolic type. The Albanese mapping $\alpha : V \longrightarrow A(V)$ is surjective and all the fibres of α are connected. Moreover the fibre space $\alpha : V \longrightarrow A(V)$ is birationally equivalent to a fibre bundle over $A(V)$ in the etale topology whose fibre and structure group are an algebraic manifold F of parabolic type and the automorphism group $\mathrm{Aut}(F)$ of F, respectively.

As a special case of Conjecture K_n, we have the following :

Conjecture B_n. Let V be an n-dimensional algebraic manifold of parabolic type. Suppose that there exists a surjective morphism $f : V \longrightarrow A$ of V onto an n-dimensional abelian variety A. Then the Albanese mapping $\alpha : V \longrightarrow A(V)$ is a modification.

In case of algebraic surfaces, Conjecture K_2 is true. Let S be a minimal algebraic surface of parabolic type. If $q(S) = 2$, then S is an abelian surface. Hence, the Albanese mapping $\alpha : S \longrightarrow A(S)$ is an isomorphism. If $q(S) = 1$, then S is a hyperelliptic surface and the Albanese mapping $\alpha : S \longrightarrow A(S)$ has the structure of an elliptic bundle over an elliptic curve $A(S)$ (see 20.15, below).

As for Conjecture B_n, we have the following partial answer.

Lemma 11.4.2. Let $f : V \longrightarrow A$ be the same as in Conjecture B_n. If the field extension $\mathbb{C}(V)/\mathbb{C}(A)$ is Galois, then Conjecture B_n holds.

Proof. Let ω be a holomorphic n-form A. Let $D = \sum n_i C_i$ be a canonical divisor of V defined by the zeros of a holomorphic n-form $f^*\omega$ on V. The support of D consists of those points at which f is not locally biholomorphic. Suppose that there exists an irreducible component, say C_1 of D such that $f(C_1)$ is of codimension one in A. Since f is Galois, there exist irreducible components $C_2, C_3, \dots, C_m$ of D such that $n_1 = n_2 = \dots = n_m$ and

$f^{-1}(f(C_1)) = \bigcup_{i=1}^{m} C_i$. From Theorem 5.13, we have

$$\kappa(V) = \kappa(D, V) \geqq \kappa(\sum_{i=1}^{m} n_i C_i, V) = \kappa(f(C_1), A) > 0,$$

since every effective divisor on an abelian variety has a positive D-dimension (see Weil [2], Théorèm 1, p.114 and Proposition 7, p.121). This is a contradiction. Hence for any irreducible component C_i of D, $f(C_i)$ is of codimension at least two. Then it is easy to show that V is birationally equivalent to a finite unramified covering of A. That is, V is birationally equivalent to an abelian variety.

Q.E.D.

The following proposition is an easy consequence of Theorem 11.3.1.

Proposition 11.4.3. Let V be an n-dimensional algebraic manifold of parabolic type. Suppose that there exists a birationally equivalent model V^* of V such that mK_{V^*} is trivial for a positive integer m. Then Conjecture K_n is true for the manifold V.

As we have already mentioned, an algebraic manifold of parabolic type may not satisfy the assumption of Proposition 11.4.3. In § 16, we shall show that for a Kummer manifold V (see Definition 16.1), Conjecture K_n is true even if the Kummer manifold V does not satisfy the assumption of Corollary 11.14.3 (see Theorem 16.7 and Remark 16.8).

(11.5) To study the Albanese mapping of an algebraic manifold of parabolic type, we consider the following conjecture due to Iitaka [3].

Conjecture C_n. Let $f : V \longrightarrow W$ be a fibre space of an n-dimensional algebraic manifold V over an algebraic manifold W. Then we have

$$\kappa(V) \geqq \kappa(W) + \kappa(V_w),$$

where $V_w = f^{-1}(w)$, $w \in W$ is a general fibre of f.

From Proposition 6.13 and Corollary 10.6, we obtain the following:

Lemma 11.5.1. If Conjecture C_n is true, then Conjecture A_n and Conjecture Q_n are true. Moreover, if Conjecture B_n and Conjecture C_n are true, then all the fibres of the Albanese mapping $\alpha : V \longrightarrow A(V)$ of an algebraic manifold of parabolic type are connected and a general fibre of the morphism α is of parabolic type.

Hence, if we shall be able to prove a canonical bundle formula for such a fibre space (see 11.8, below), it may be possible to prove Conjecture K_n (see 11.9).

Theorem 11.5.2. Conjecture C_2 is true.

Proof. We use freely the results on classification theory of surfaces (see §20, below). We assume that genera of a general fibre of the fibre space $f : V \longrightarrow W$ and W are positive since otherwise C_2 is trivially true.

Case 1). $\kappa(V) = 0$. By classification theory of surfaces, we have

$$2 \geq q(V) \geq q(W) \geq 1.$$

If $q(V) = 2$, then V is birationally equivalent to an abelian variety. Since there does not exist a morphism from an abelian variety onto a curves of genus $g \geq 2$, $q(W) = 1$ in this case. From Poincaré's reducibility theorem, we infer that a general fibre of the morphism f is an elliptic curve. Hence, in this case, C_2 is true.
If $q(V) = 1$, then V is birationally equivalent to a hyperelliptic surface and W is an elliptic curve. Since a hyperelliptic surface has a finite unramified covering which is a product of two elliptic curves, a general fibre of f is also an elliptic curve.

Therefore, C_2 holds.

Case 2). $\kappa(V) = 1$. The surface V is an elliptic surface. If a general fibre of f is an elliptic curve, then C_2 is true by virtue of the canonical bundle formula for elliptic surfaces (see Kodaira [3], I, p.772 and 20.13.1, below). Suppose that a general fibre of f is a curve of genus $g \geqq 2$. As V is an elliptic surface, there exists a fibre space $h : V \longrightarrow C$ over a curve C such that a general fibre $V_a = h^{-1}(a)$, $a \in C$ of the morphism h is an elliptic curve. Since a general fibre V_a of the morphism h is not contained in a fibre of f, we have $f(V_a) = W$. Hence W is an elliptic curve (we have assumed that W has a positive genus). Therefore, C_2 is true.

Case 3) $\kappa(V) = 2$. In this case, C_2 is trivially true. Q.E.D.

In the above proof, we used the deep results on classification theory of surfaces. The proof without using the classification theory is not yet known. Probably the study of fibre spaces of curves of genus two due to Namikawa and Ueno [1], [2] and Ueno [8] will give an information to this direction (for example, Ueno [8] has shown the formula for K_S^2 of a fibre space of curves of genus two over a curve, which is deeper than the validity of C_2).

As for higher dimensional fibre spaces, only a few results concerning Conjecture C_n are known.

<u>Theorem 11.5.3</u>. C_3 is true for a fibre space of principally polarized abelian surfaces over a curve and an elliptic threefold which has locally meromorphic sections (in the complex topology) at any point of the base surface.

By elliptic threefold V, we mean a fibre space $f : V \longrightarrow W$ of a threefold V onto a surface W such that a general fibre of f is an elliptic curve. For the definition of a fibre space of principally polarized abelian surfaces, see Ueno [1], I, Definition 1.3, p.42. The above theorem 11.5.3 is a corollary of canonical bundle formulae for such fibre spaces given by Ueno [1], III and Ueno [3], I (see also 11.8 below).

Theorem 11.5.4. If the fibre space $f : V \longrightarrow W$ of algebraic manifolds has a structure of an analytic fibre bundle over W whose fibre and structure group is an algebraic manifold F and the automorphism group Aut(F) of F, respectively, then C_n is true.

This will be proved in §15 below.

Conjecture C_n is also deeply related to the structure of algebraic varieties of elliptic type. Before we shall discuss it, we shall provide an interesting problem on algebraic manifolds of parabolic type.

Problem 11.6. Does there exist an algebraic manifold V of parabolic type such that $q^*(V) = 0$ and the fundament group $\pi_1(V)$ of V is an infinite group ?

For every known example of algebraic manifold V of parabolic type with $q^*(V) = 0$, $\pi_1(V)$ is a finite group (for surfaces, see 20.14 and 20.16).

(11.7) Now we shall consider a complex manifold V of elliptic type. If the algebraic dimension a(V) is zero, V is of elliptic or of parabolic type. Such a manifold will be studied in §13. Unirational varieties (see Example 6.6,2)) and rational varieties are of elliptic type.

Suppose that V is an algebraic manifold of elliptic type. If Conjecture C_n is true, then a general fibre of the fibre space $\beta : V \longrightarrow W$ associated with the Albanese mapping $\alpha : V \longrightarrow A(V)$ is an algebraic variety of elliptic type by virtue of Corollary 10.6 (in the proof of Theorem 16.9, we use this fact). Then the study of algebraic manifolds of elliptic type is reduced to

1) the study of algebraic manifolds of elliptic type of Albanese dimension zero ;

2) the study of fibre spaces whose general fibre is of elliptic type.

To find a criterion of rationality is one of the most important problem concerning the first problem. For algebraic threefolds, it has been studied for a long time (see Roth [1]). Recently it is shown that non-singular cubic threefolds are not rational (cubic threefolds are always unirational) (see Clemens and Griffith [1], Tjurin [1] and Murre [1]) and non-singular quartic threefolds are not rational (some of them are unirational, but it is not known whether all non-singular quartic threefolds are unirational or not) (see Iskovskih and Manin [1]). Artin and Mumford [1] has constructed an algebraic threefold which is not rational but unirational.

There are many problems on algebraic manifolds of elliptic type which we should consider . Here we only give the following interesting problem.

<u>Problem 11.7.1</u>. Does there exist an algebraic manifold V of elliptic type such that V is not unirational and $g_k(V) = 0$, $k = 1, 2, \dots, \dim V$?

For example, let $F^{(3)}$ be a non-singular model of a quotient variety $E \times E \times E/G$ where E is an elliptic curve with fundamental periods $\{1, \sqrt{-1}\}$ and G is the cyclic group of order four of analytic automorphisms of $E \times E \times E$ generated by an automorphism

$$g : (z_1, z_2, z_3) \longmapsto (\sqrt{-1}z_1, \sqrt{-1}z_2, \sqrt{-1}z_3).$$

From Proposition 9.24, we infer that $g_k(V) = 0$, $k = 1, 2, 3$. Moreover, using the canonical resolution of a cyclic quotient singularity given in 16.10, we can prove that $\kappa(V) = -\infty$. It is not known whether $F^{(3)}$ is unirational or not (see Example 16.15, below).

We also refer the reader to Fujita [1] (see also 19.7 below) Iitaka [6] and Martynov [1].

(11.8) Finally we shall consider fibre spaces whose general fibres are of parabolic type. One of the most important problem is to give canonical bundle formulae for such fibre spaces. There are a few cases in which we know the answer. For an elliptic surface (see Definition 20.6.2) and 20.13), Kodaira [3], I has shown the canonical bundle formula (see 20.13.1, below). Ueno [3] has generalized the formula in the following way.

Theorem 11.8.1. Let $f : V \longrightarrow W$ be a surjective morphism of a threefold V onto a surface W such that general fibres are elliptic curves. Suppose that the image of all points on V where f is not of maximal rank is a divisor with normally crossings and that f has locally meromorphic sections (in the complex topology) at any point of W. Then a twelfth canonical divisor $12K(V)$ has a form

$$f^*(12K(W) + F) + G,$$

where G is an effective divisor on V which does not come from a divisor on W and F is a divisor on W written is the form

$$F = \sum_b b\, S_{I_b} + \sum_b (6 + b)\, S_{I_b^*} + 2S_{II} + 10S_{II^*} + $$
$$+ 3S_{III} + 9S_{III^*} + 4S_{IV} + 8S_{IV^*} ,$$

such that $S_{(*)} = \sum S_j$ and S_j is a non-singular curve over which the singular fibre is of type Kod(*).

The singular fibre of type Kod(*) over a curve S_j is defined as follows. Let D be a small disk in the surface W such that D intersect the curve S_j transversally at one point p which is the origin of the disk D. We restrict the fibre space $f : V \longrightarrow W$ to D and obtain a fibre space $g : V_D \longrightarrow D$. If the Picard-Lefschetz transformation of this fibre spaces $g : V_D \longrightarrow D$ along a circle rounding the origin once counterclockwise is $SL(2, \mathbb{Z})$-conjugate to a matrix of type (*) given in Kodaira [2], II, Table I, p.604, then the singular fibre over the curve S_j is called of type Kod (*). For the detailed discussion and the proof of Theorem 11.18.1, see Ueno [3], I, §4 - §6. We remark that the above theorem and the proof given in Ueno [3] are also valid for a higher dimensional elliptic fibre space which has locally meromorphic sections. The divisor G is a fixed component of $|12K(V)|$.

We also remark that the canonical bundle formula for a fibre space of principally polarized abelian surfaces over a curve has been obtained by Ueno [1], III.

(11.9) Now we shall show that the above canonical bundle formula is deeply related to Conjecture K_n. Let V be an n-dimensional algebraic manifold of parabolic type with $q(V) = n - 1$. If the Albanese mapping $\alpha : V \longrightarrow A(V)$ is surjective with connected fibres and if Conjecture C_n is true, then general fibres are elliptic curves. Suppose, moreover, that the fibre space $\alpha : V \longrightarrow A(V)$ has locally meromorphic sections (in the complex topology) at any point of $A(V)$. Let S be an image of the morphism α of all points in V at which the morphism α is not of maximal rank. If S is not a divisor with normally crossings, by finite succession of monoidal transformations with non-singular centers, we obtain an algebraic manifold W and a surjective morphism $g : W \longrightarrow A(V)$ such that the strict transform $\tilde{S}$ of S is a divisor with normally crossings. Let V^* be a non-singular model of $V \underset{A(V)}{\times} W$ constructed in Theorem 2.12. We have a natural surjective morphism $f : V^* \longrightarrow W$ and the divisor $\tilde{S}$ coincides with the image of the morphism f of those points in V^* at which f is not of maximal rank. The fibre space $f : V^* \longrightarrow W$ is birationally equivalent to the fibre space $\alpha : V \longrightarrow A(V)$. Applying Theorem 11.18.1 and Theorem 5.13 to the fibre space $f : V^* \longrightarrow W$, we obtain

$$\kappa(V) = \kappa(V^*) \geqq \kappa(f^*(12K(W) + F), V^*) = \kappa(12K(W) + F, W) = \kappa(F,W),$$

since the canonical divisor $K(W)$ consists of exceptional varieties appearing in the monoidal transformations. Let F^* be a divisor on the Albanese variety $A(V)$ such that $F = g^*F^* + D$ where D is an effective divisor which appears in the monoidal transformations. Then we have

$$\kappa(F, W) = \kappa(g^*F^*, W) = \kappa(F^*, A(V)).$$

From the theory of theta functions (see, for example, Weil [2], Théorèm 1, p.114 and Proposition 7, p.121), we infer that, if F^* is an effective divisor, then $\kappa(F^*, A(V)) > 0$. On the other hand, by our assumption, we have

$$0 = \kappa(V) \geqq \kappa(F^*, A(V)).$$

Hence F^* must be zero. From the theory of elliptic fibre spaces (see Kodaira [2], II, Kawai [2] and Ueno [3], I), we infer readily that V is birationally equivalent to a fibre bundle in the etale topology over A(V) whose fibre and structure group are an elliptic curve E and the automorphism group Aut(E), respectively. This proves Conjecture K_n in our case.

Chapter V

Algebraic reductions of complex varieties and complex manifolds of algebraic dimension zero

In this chapter, we shall study the algebraic reductions of complex varieties (Definition 3.3) and the Albanese mappings of complex manifolds of algebraic dimension zero. The fibre spaces introduced by these two morphisms have similar properties. This is the reason why we have postponed the study of algebraic reductions up to now.

In §12, first we shall study the structure of general fibres of algebraic reductions. The study of algebraic reductions has been begun by Chow and Kodaira [1] and Kodaira [2]. Chow and Kodaira [1] has shown that, for a surface S, if $a(S) = 2$, then S is projective. Kodaira [2], I has shown that, for a surface S, if $a(S) = 1$, then there exists a surjective morphism $f : S \longrightarrow C$ of S onto a curve C such that a general fibre of f is an elliptic curve. Kawai [1], I has shown that, for a threefold M, if $a(M) = 2$, then a general fibre of the algebraic reduction of M is an elliptic curve. Hironaka [3] has pointed out that Kawai's argument can be simplified using the projective fibre space associated with a coherent sheaf. Following Hironaka's idea, Kawai [2], II has studied the structure of fibres of algebraic reductions of threefolds of algebraic dimension one. Following their arguments, we shall prove that the Kodaira dimensions of general fibres of algebraic reduction are always non-positive (see Corollary 12.2). If general fibres of algebraic reductions are curves or surfaces, using classification theory of curves and surfaces, we obtain more precise informations about the general fibres. Finally we shall give the proof of Theorem 3.8 announced in §3.

In §13, we shall study the structure of complex manifolds of algebraic dimension zero using the Albanese mappings. The study of such complex manifolds has been begun by Kodaira [2], I. He has studied Kähler surfaces of algebraic dimension zero using the Albanese mappings. Kawai [1], II has studied Kähler threefolds of algebraic dimension zero by the similar method. By virtue of Corollary 10.6, we can remove the restriction that complex manifolds are Kähler. We shall show that the Albanese mapping $\alpha : M \longrightarrow A(M)$ of a complex manifold M of algebraic dimension zero is surjective with connected fibres (see Lemma 13.1 and Lemma 13.6). If general fibres of the Albanese mapping α are curves or surfaces, we obtain more precise informations on the general fibres. We note that the existence of moduli spaces of curves and surfaces as algebraic spaces, and hyperbolic analysis due to Kobayashi and other mathematicians play the essential role in our theory.

§12. Algebraic reductions of complex varieties

In this section we shall study the structure of fibres of an algebraic reduction of a complex variety (Definition 3.3). We have already shown that all the fibres of an algebraic reduction are connected (Proposition 3.4). We begin with the following theorem.

Theorem 12.1. Let $\Psi^* : M^* \longrightarrow V$ be an algebraic reduction of a complex variety M. For any Cartier divisor D on M^*, there exists a dense subset U of V such that the fibre $M^*_u = \Psi^{*-1}(u)$ over $u \in U$ is non-singular and we have

$$\kappa(D_u, M^*_u) \leqq 0 ,$$

for any point $u \in U$ where D_u is the restriction of the divisor D to the fibre M^*_u.

Proof. For any positive integer m, we set

$$\underline{L}_m = \Psi^*_*(\underline{O}(mD)) .$$

Since $\underline{L}_m$ is a coherent sheaf by Theorem 1.4, there is a Zariski open set $U_1^{(m)}$ of V such that the restriction $\underline{L}_m|U_1^{(m)}$ is either a locally free sheaf of rank n or the zero sheaf. By Corollary 1.8, there is a Zariski open set U_2 of V such that $M^*_u = \Psi^{*-1}(u)$ is non-singular for any point $u \in U_2$. Assume that there is a positive integer m such that $\underline{L}_m|U_3^{(m)}$, $U_3^{(m)} = U_1^{(m)} \cap U_2$, is a locally free sheaf of rank $n \geqq 2$. By Theorem 1.4,3), for a Zariski open subset $U^{(m)}$ of $U_3^{(m)}$,

$$\dim_{\mathbb{C}} H^0(M^*_u, \underline{O}(mD_u)) = n \geqq 2, \quad u \in U^{(m)} .$$

Hence, $\kappa(D_u, M^*_u) \geqq 1$. Consider the projective fibre space $g : \mathbb{P}(\underline{L}_m) \longrightarrow V$ associated with the coherent sheaf $\underline{L}_m$. As we have seen in 2.10, there is a commutative diagram

$$\begin{array}{ccc} M^* & \xrightarrow{\quad h \quad} & \mathbb{P}(\underline{L}_m) \\ & {\scriptstyle f} \searrow \quad \swarrow {\scriptstyle g} & \\ & V & \end{array}$$

where h is a meromorphic mapping and the restriction

$$h_u : M_u^* \longrightarrow \mathbb{P}(\underline{L}_m)$$

is nothing other than the mapping

$$\Phi_{mD_u} : M_u^* \longrightarrow W_m$$

where $u \in U^{(m)}$. Hence $\dim h(M^*) > \dim V$. On the other hand, since $\underline{L}_m$ is coherent, by GAGA (Theorem 1.3), there is an algebraic coherent sheaf $\underline{L}_m{}^{alg}$ on V^{alg} with

$(\underline{L}_m^{alg})^{an} \xrightarrow{\sim} \underline{L}_m$. It follows that $\mathbb{P}(\underline{L}_m^{alg})^{an} \xrightarrow{\sim} \mathbb{P}(\underline{L}_m)$, $\mathbb{P}(\underline{L}_m)$ is an algebraic variety and $h(M^*)$ is an algebraic subvariety of $\mathbb{P}(\underline{L}_m)$. Calculating algebraic dimensions, we find

$$a(M^*) \geqq a(h(M^*)) = \dim h(M^*) > \dim V = a(M^*) .$$

This is a contradiction. Hence $\underline{L}_m|_{U^{(m)}}$ is an invertible sheaf or the zero sheaf. We set $U = \bigcap_{m \geq 1} U^{(m)}$. Since V is a Baire space, U is a dense set. Since $\underline{L}_m|_U$ is zero or locally free of rank one for any positive integer m, we conclude that $\kappa(D_u, M_u^*) \leqq 0$.

Q.E.D.

From the above argument, we infer readily the following :

<u>Proposition 12.2</u>. Let $f : M \longrightarrow V$ be a fibre space of complex varieties. Suppose that V is a Moishezon variety and that there exist a Cartier divisor D on M and an open set U in V such that $M_u = f^{-1}(u)$ is non-singular and $\kappa(D_u, M_u) = m > 0$ for any point $u \in U$ where D_u is the restriction of the divisor D to M_u.

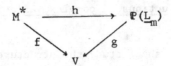

Then we have

$$a(M) \geqq \dim V + m .$$

<u>Corollary 12.3</u>. Let $\Psi^* : M^* \longrightarrow V$ be the algebraic reduction of a complex variety M. There exists a dense subset U of V such that $\kappa(M_u^*) \leqq 0$ for any $u \in U$.

It is not known whether we can choose the above dense set U as Zariski open set (see also Remark 7.5 and Remark 7.6). However, if $a(M) \leqq \dim M - 2$, this is the case and we have the following theorem.

<u>Theorem 12.4</u>. Let $\Psi^* : M^* \longrightarrow V$ be the algebraic reduction of M. If $a(M) \leqq \dim M - 2$, there exists a Zariski open set U of V which satisfies the following properties.

1) If $a(M) = \dim M - 1$, then $M_u^* = \Psi^{*-1}(u)$ is an elliptic curve for any point $u \in U$.

2) If $a(M) = \dim M - 2$, then $\kappa(M_u^*) \leqq 0$ for any point $u \in U$. Moreover M_u^* is not $\mathbb{P}^2$ for any $u \in U$.

<u>Proof</u>. By Corollary 1.8, there exists a Zariski open set U such that Ψ^* is of maximal rank at any point of $\Psi^{*-1}(U)$. As $a(M) \leqq \dim M-2$, $\dim M_u^* \leqq 2$ for any $u \in U$. Hence, by deformation invariance of the Kodaira dimension and pluri-genera of surfaces and curves (see Theorem 20.11, below), we conclude that $\kappa(M_u^*) \leqq 0$ for any $u \in U$ (see the proof of Theorem 12.1).

Suppose that M_u^* is $\mathbb{P}^1$ or $\mathbb{P}^2$ for a point $u \in U$. Then for any point $u \in U$, M_u^* is $\mathbb{P}^1$ or $\mathbb{P}^2$ (see for example Kodaira and Spencer [1], II, Theorem 20.1, p.459). Hence the sheaf $\underline{F} = \Psi^*_*(\underline{O}(-K(M)))$ is a locally free sheaf of rank $n \geqq 3$ on U. By the similar arguments as those in the proof of Theorem 12.1, using $\underline{F}$ instead of $\underline{L}_m$, we have a contradiction. Q.E.D.

Remark 12.5. By the classification of surfaces (see §20), a general fibre of the algebraic reduction in Theorem 12,4,2) is one of the following surfaces.

(1) K 3 surface. (2) complex torus. (3) hyperelliptic surface. (4) Enriques surface. (5) elliptic surface with a trivial cononical bundle. (6) surface of class VII. (7) rational surface. (8) ruled surface.

It seems very plausible that a ruled surface of genus $g \geqq 2$ (see Definition 20.6,1)) does not appear as a general fibre of the algebraic reduction of M. If $\dim M = 3$, Kawai [1], II has shown that this is the case. His main idea is as follows. Let $\Psi^* : M^* \longrightarrow \Delta$ be an algebraic reduction of a three-dimensional complex manifold M with $a(M) = 1$. Suppose that general fibres of the algebraic reduction Ψ^* are ruled surfaces of genus $g \geqq 2$. There exists a finite number of points $\{a_1, a_2, \ldots, a_m\}$ on the curve Δ such that if we set $\Delta' = \Delta - \{a_1, a_2, \ldots, a_m\}$, $M' = \Psi^{*-1}(\Delta')$ and $\Psi' = \Psi^*|_{M'}$, Ψ' is of maximal rank at any point of M'. Since $\Psi' : M' \longrightarrow \Delta'$ is a differentiable fibre bundle, using a period mapping of the universal covering manifold of Δ' into the Siegel upper half plane of degree g, we can construct a fibre space $\varphi' : S' \longrightarrow \Delta'$ of curves of genus g such that for any point $x \in \Delta'$, the fibre $S_x = \varphi'^{-1}(x)$ is a curve of genus g which is the image of the Albanese mapping $\alpha : S_x \longrightarrow A(S_x)$. Moreover, there exists a surjective morphism $\pi' : M' \longrightarrow S'$ such that the restriction $\pi_x = \pi'|_{M_x} : M_x = \Psi^{*-1}(x) \longrightarrow$

S_x of π' to the fibre M_x is the Albanese mapping for any point $x \in \Delta'$ (see the last part of the proof of Theorem 13.11). The above process is always possible even if $\dim M \geqq 4$. Kawai has shown that, if $\dim M = 3$, then the fibre space $\varphi' : S' \longrightarrow \Delta'$ can be extended to a fibre space $\varphi : S \longrightarrow \Delta$ in such a way that the morphism $\pi' : M' \longrightarrow W'$ can be extended to a meromorphic mapping $\pi : M^* \longrightarrow S$ of M onto S (if $\dim M \geqq 4$, we can always construct a fibre space $\varphi : S \longrightarrow \Delta$ which is an extension of $\varphi' : S' \longrightarrow \Delta'$, but it is not known whether the morphism $\pi' : M' \longrightarrow S'$ can be extended to a meromorphic mapping $\pi : M^* \longrightarrow S$ of M^* onto S or not). Since general fibres of the fibre space $\varphi : S \longrightarrow \Delta$ are curves of genus $g \geqq 2$, by Corollary 12. 3, we have $a(S) = 2$. Hence, $a(M^*) \geqq 2$. This is a contradiction.

It is also interesting to know whether a hyperelliptic surface, an Enriques surface and a rational surface appear as a general fibre (see also Remark 13.12, below).

<u>Example 12.6</u>. 1) Let S be a K3 surface, a complex torus or a surface of class VII with $a(S) = 0$ and let V be a projective manifold. We set $M = S \times V$. The projection $p : M \longrightarrow V$ of M to the second factor is the algebraic reduction of M.

2) (Kawai [1], II) Let G be a group of analytic automorphisms of $\mathbb{C} \times \mathbb{P}^1$ generated by automorphisms σ, τ defined by

$$\sigma : (z, x) \longmapsto (z + 1, ax + b),$$

$$\tau : (z, x) \longmapsto (z + \sqrt{-1}, cx + d), \quad ac \neq 0,$$

where z is a global coordinate of $\mathbb{C}$, x is a nonhomogeneous coordinate of x and a, b, c, d are constants. The quotient space. $S = \mathbb{C} \times \mathbb{P}^1/G$ is a ruled surface of genus one. Let E be an elliptic curve with fundamental periods $\{1, \sqrt{-1}\}$. The projection of $\mathbb{C} \times \mathbb{P}^1$

onto the first factor induces a surjective morphism

$$\begin{array}{ccc} \alpha : S & \longrightarrow & E \\ \Psi & & \Psi \\ [z, x] & \longmapsto & [z], \end{array}$$

which is nothing other than the Albanese mapping of S. Let $\pi : T \longrightarrow R$ be a principle E-bundle over a non-singular curve R such that $a(T) = 1$. Since translations of $\mathbb{C}$ commute with σ and τ, the elliptic curve E operates on S. Hence we have a fibre bundle $f : M \longrightarrow R$ associated with $\pi : T \longrightarrow R$ whose fibre is the surface S. We shall show that, if the constants a, b, c, d are general, $a(M) = 1$ and f is the algebraic reduction of M. Suppose the contrary Since $a(T) = 1$, we have $a(M) = 2$. Hence, by Theorem 12.4, 1), we have an algebraic reduction $\Psi^* : M^* \longrightarrow V$ such that general fibres are elliptic curves. Let S^* be the strict transform of a fibre of $f : M \longrightarrow R$, that is, the surface S, to M^*. Then $\Psi^*(S^*)$ is a curve on V, because, if $\Psi^*(S^*) = V$, then by Proposition 12.2, we have $a(M^*) = 3$. It follows that S^* is an elliptic surface. Hence S is also an elliptic surface. But it is easy to show that for general constants a, b, c, d, S has not a structure of an elliptic surface.

3) The example given in Example 13.13, 3) below with an abelian variety T instead of a torus of algebraic dimension zero gives an example of a complex manifold such that general fibres of an algebraic reduction are elliptic surfaces with a trivial canonical bundle.

Before we shall prove Theorem 3.8, we give one more remark.

<u>Remark 12.7</u>. It is interesting to know when we can take M as M^* for an algebraic reduction $\Psi^* : M^* \longrightarrow V$ of a complex manifold M. For example, if $a(M) = 1$ and the genus of the function field

$\mathbb{C}(M)$ is positive, then we can always take M as M^*. This comes from the fact that any meromorphic mapping of a complex manifold onto a curve of genus $g \geqq 1$ is holomorphic (we can use the similar argument as that in the proof of Lemma 9.11). But if $\mathbb{C}(M)$ is a purely transcendental extension of $\mathbb{C}$, we cannot necessarily take M as M^* (see Lemma 18.1.3 and Remark 18.1.4, below).

Finally we shall prove Theorem 3.8.

(12.8) <u>Proof of Theorem 3.8.1)</u>. First we shall prove the inequality

(12.8.1) $a(M) \leq a(N) + \dim f$,

by induction on $\dim M$. Clearly, taking non-singular models of M and N, we can assume that M and N are non-singular. As f is surjective, $\mathbb{C}(M) \supset \mathbb{C}(N)$. If $a(M) = a(N)$, the assertion is true. Hence, we can assume that $a(M) > a(N)$. In that case, there is an element $g \in \mathbb{C}(M)$ which does not belong to $\mathbb{C}(N)$. g defines a meromorphic mapping

$$\begin{array}{ccc} g : M & \longrightarrow & \mathbb{P}^1 \\ \cup\!\!\!\!\cup & & \cup\!\!\!\!\cup \\ z & \longmapsto & (1 : g(z)) \end{array} .$$

Let $p : M^* \longrightarrow M$ be a modification where M^* is smooth and $g^* = g \circ p : M^* \longrightarrow \mathbb{P}^1$ is a morphism. Consider the Stein factorization

$$\begin{array}{ccc} M^* & \xrightarrow{\quad g^* \quad} & \mathbb{P}^1 \\ & {\scriptstyle g_1} \searrow \quad \nearrow {\scriptstyle g_2} & \\ & C & \end{array}$$

of g^*. As C is a curve, C contains a finite number of points $y_1, y_2, \ldots, y_\ell$ such that $M^*_y = g_1^{-1}(y)$, $y \in C' = C - \{y_1, y_2, \ldots, y_\ell\}$ is irreducible and non-singular. Suppose that, for each point $y \in C'$, $f(M^*_y) \subsetneqq N$. We can find, then, a point α of $\mathbb{P}^1$ such that, if we set $f^* = f \circ p$, we have

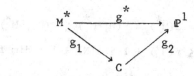

$$X = f^*(\bigcup_{x \in g_2^{-1}(\alpha)} M_x^*) \subsetneq N .$$

Hence, for any point $u \in N - X$, $g_u^* = g^*|_{f^{*-1}(u)} : f^{*-1}(u) \longrightarrow \mathbb{P}^1$ does not take a value α. Hence we can consider g_u^* as a holomorphic function on $f^{*-1}(u)$. As $f^{*-1}(u)$ is compact, g_u^* is a constant. This means that the meromorphic function g is constant on the fibre $f^{-1}(u)$, $u \in N - X$. Since N is normal, g comes from a meromorphic function on N (see the proof of Proposition 1.13). This contradicts the assumption that $g \notin \mathbb{C}(N)$. Hence there is a point $y \in C'$ such that $f(M_y^*) = N$. From our induction hypothesis (if M is a curve, Theorem 3.8,1) is clearly true), we obtain,

$$a(M_y^*) \leqq a(N) + \dim f_y^* ,$$

where we set $f_y^* = f^*|_{M_y^*}$. Next, if H is a hyperplane divisor on V and $D = (f^*)^* H$ is the pull back of H to M, we can easily show that $a(M^*) = \kappa(D, M^*)$. Applying Theorem 5.11 to the morphism $g^* : M^* \longrightarrow C$, we obtain the inequality

$$\kappa(D, M^*) \leqq \kappa(D_y, M_y^*) + 1 .$$

Moreover, by Lemma 5.5, we have $\kappa(D_y, M_y^*) \leqq a(M_y^*)$. Combining these inequalities, we have

$$\begin{aligned} a(M) = a(M^*) &= \kappa(D, M^*) \leqq \kappa(D_y, M_y^*) + 1 \\ &\leqq a(M_y^*) + 1 \leqq a(N) + 1 + \dim f_y^* \\ &= a(N) + \dim f^* \\ &= a(N) + \dim f . \end{aligned}$$

Next we shall show that there exists a nowhere dense analytic subset N_1 of N such that

$$a(M) \leqq a(N) + a(M_y) ,$$

for any $y \in N - N_1$. We can assume that $\Psi_M : M \longrightarrow V$ and $\Psi_N : N \longrightarrow W$ are algebraic reductions of M and N, respectively and there exists a morphism $f_0 : V \longrightarrow W$ such that $\Psi_N \circ f = f_0 \circ \Psi_M$. (If necessary, we can replace M, N, V and W by their bimeromorphically equivalent models). Let L be the image variety of the morphisms

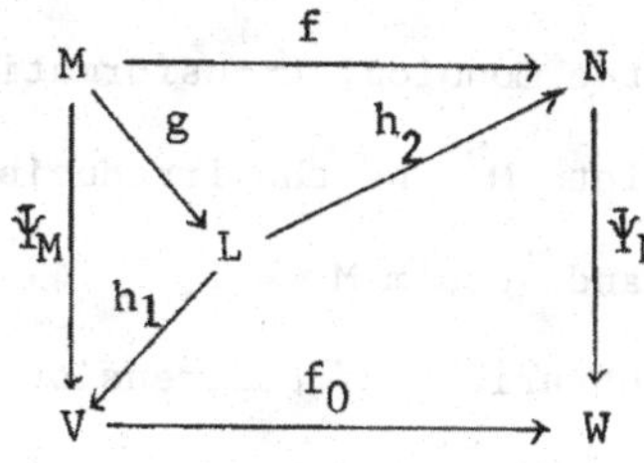

$g = \Psi_M \times f : M \longrightarrow V \underset{W}{\times} N$. $h_1 : L \longrightarrow V$ and $h_2 : L \longrightarrow N$ are the natural morphisms induced by the projections of $V \underset{W}{\times} N$ onto the first and the second factor, respectively.

Then the morphism h_1 and h_2 are surjective and we have

$$a(M) = a(L) = a(V) .$$

Let N_1 be a nowhere dense analytic subset of N such that for any $y \in N - N_1$, the fibres $M_y = f^{-1}(y)$ and $L_y = h_2^{-1}(y)$ are irreducible, and $\dim M_y$ and $\dim L_y$ are constant on $N - N_1$. Since h_1 induces an isomorphism between L_y and $\Psi_M(M_y)$, and V is algebraic, we obtain

$$a(L_y) = \dim L_y, \quad a(L_y) \leqq a(M_y) .$$

Applying the above result and the inequality 12.8.1 to the fibre space $h_2 : L \longrightarrow N$, we obtain inequalities

$$a(M) = a(L) \leqq a(N) + \dim L_y = a(N) + a(L_y)$$
$$\leqq a(N) + a(M_y). \qquad \text{Q.E.D.}$$

The inequality

$$a(M) \leqq a(N) + \dim f$$

was proved by Ueno and Iitaka (see Iitaka [2]). Ma. Kato and Fujita has shown independently the inequality

$$a(M) \leqq a(N) + a(M_y).$$

(12.9) Proof of Theorem 3.8.2). Here we can use the arguments of Moishezon [1], Chap. I, Theorem 3. For reader's convenience, we shall give the proof.

First, we reduce to the case where codim $M = 1$. Suppose that codim $M \geqq 2$. Let $(N, f), f : N^* \longrightarrow N$ be the monoidal transformation of N with center M (Definition 2.11) and let M^* be the irreducible component of $f^{-1}(M)$ such that $f(M^*) = M$ and codim $M^* = 1$. Assuming that the assertion is true for a subvariety of codimension one in N^*, we see that $a(N^*) \leqq a(M^*) + 1$. However, the first part of our theorem implies that

$$a(M^*) \leqq a(M) + \text{codim } M - 1 .$$

Hence, as a consequence of these two inequalities, we have

$$a(N) = a(N^*) \leqq a(M) + \text{codim } M .$$

Therefore, we can assume that codim $M = 1$.

Let $\iota : N^* \longrightarrow N$ be the normalization of N and $\iota^{-1}(M) = M^*$. As ι is a finite morphism, it follows that $a(N) = a(N^*)$ and that $a(M) = a(M^*)$. Thus, we can also assume that N is normal. We set

$$L(N, M) = \{ f \in \mathbb{C}(N) \mid \text{The pole of } f \text{ does not contain } M\},$$

and

$$I(N, M) = \{ f \in L(N, M) \mid f_{|M} = 0 \} .$$

$F = L(N, M)/I(N, M)$ can be imbedded into $\mathbb{C}(M)$.

Next, we proceed under the following assumption :

T) The ring $L(N, M)$ contains a field $\tilde{F}$ which is isomorphic to F under the natural homomorphism

$$L(N, M) \longrightarrow L(N, M)/I(N, M) = F .$$

Let $y_1, y_2, \ldots, y_s$ be algebraically independent elements in F where $s = \mathrm{tr.deg.}_{\mathbb{C}} F$. There are algebraically independent $x_1, x_2, \ldots, x_s$ in $\tilde{F}$ corresponding to the $y_1, y_2, \ldots, y_s$. Consider an element t of $I(N, M)$ which has a zero of the least multiplicity on M. Then, $x_1, x_2, \ldots, x_s, t$ are algebraically independent over $\mathbb{C}$. Now suppose that $s < a(N) - 1$. Then there exists a meromorphic function g contained in $\mathbb{C}(N)$ such that $g, x_1, x_2, \ldots, x_s, t$ are algebraically independent over $\mathbb{C}$. If g is not contained in $L(N, M)$, then we have $1/g \in L(N, M)$. Hence we suppose that $g \in L(N, M)$. As N is a normal variety, there is a coordinate neighbourhood U with coordinates $(z_1, z_2, \ldots, z_n)$ in the non-singular part of N such that $U \cap M \neq \phi$. Since $g, x_1, x_2, \ldots, x_s, t$ are algebraically independent, it must be the case that

$$dg \wedge dx_1 \wedge dx_2 \wedge \cdots \wedge dx_s \wedge dt \neq 0$$

on the non-singular part of N (see Remmert [1]: algebraic dependence and analytic dependence of meromorphic functions on a complex variety are equivalent). Hence on U we can assume that

$$J = \det \begin{pmatrix} \frac{\partial g}{\partial z_1} & \frac{\partial x_1}{\partial z_1} & \cdots & \frac{\partial x_s}{\partial z_1} & \frac{\partial t}{\partial z_1} \\ \frac{\partial g}{\partial z_2} & \frac{\partial x_1}{\partial z_2} & \cdots & \frac{\partial x_s}{\partial z_2} & \frac{\partial t}{\partial z_2} \\ \cdots & \cdots & \cdots & \cdots & \cdots \\ \cdots & \cdots & \cdots & \cdots & \cdots \\ \frac{\partial g}{\partial z_{s+2}} & \frac{\partial x_1}{\partial z_{s+2}} & \cdots & \frac{\partial x_s}{\partial z_{s+2}} & \frac{\partial t}{\partial z_{s+2}} \end{pmatrix} \neq 0 .$$

J is a meromorphic function on U which has no pole on $U \cap M$. Let r be the largest integer such that J/t^r has no pole on $U \cap M$. The assumption T) implies that there is an expansion

$$g = \sum_{i=0}^{r+1} g_i t^i + t^{r+2} h$$

where $g_i \in \tilde{F}$ and $h \in L(N, M)$. As $\sum_{i=0}^{r+1} g_i t^i, x_1, \ldots, x_s, t$ are algebraically dependent, they are also analytically dependent in U. Also we have

$$\frac{\partial(t^{r+2}h)}{\partial z_k} = t^{r+1}h_k, \quad k = 1,2,\ldots,s+2,$$

where h_k is meromorphic on U and has no pole on $U \cap M$.
Calculating J, we find

$$J = t^{r+1}.\det \begin{pmatrix} h_1, \frac{\partial x_1}{\partial z_1}, \ldots, \frac{\partial x_s}{\partial z_1}, \frac{\partial t}{\partial z_1} \\ h_2, \frac{\partial x_1}{\partial z_2}, \ldots, \frac{\partial x_s}{\partial z_2}, \frac{\partial t}{\partial z_2} \\ \cdots\cdots\cdots\cdots\cdots\cdots \\ h_s, \frac{\partial x_1}{\partial z_{s+2}}, \ldots, \frac{\partial x_s}{\partial z_{s+2}}, \frac{\partial t}{\partial z_{s+2}} \end{pmatrix}.$$

Hence J/t^{r+1} has no pole on $U \cap M$. This contradict the assumption about r. Therefore, we obtain the desired conclusion

$$a(M) \geqq s \geqq a(N) - 1.$$

Finally, we shall show that it is enough to consider the case where the assumption T) is valid. $y_1, y_2, \ldots, y_s$ will be the same as above and y_{s+1} will denote the primitive element of the extension $F/\mathbb{C}(y_1, \ldots, y_s)$ having $g(X) = 0$ as a minimal equation. Let $\tilde{g}(X_0, X_1)$ be the homogeneous polynomial corresponding to the polynomial $g(X)$. Suppose that, for elements $x_i \in \mathbb{C}(N)$, $i = 1,2,\ldots,s$, we have $\tau_0(x_i) = y_i$, $i = 1,2,\ldots,s$, where τ_0 is the natural homomorphism $L(N, M) \longrightarrow F = L(N, M)/I(N, M)$. Then, replacing y_i by x_i, we obtain a homogeneous polynomial $\tilde{g}(X_0, X_1)$ with coefficients in $L(N, M)$. Let D be the analytic subset of N which is the union of the supports of the poles of the coefficients of the polynomial $\tilde{g}(X_0, X_1)$, the singular locus of M and the singular locus of N. From our choice of the $x_1, x_2, \ldots, x_s$ and the normality of M, it follows that $D \not\supset M$. Let N' be the analytic subset of the product $(N - D) \times \mathbb{P}^1$ which is defined by the equation

$$\tilde{g}(\xi_0, \xi_1) = 0$$

where $(\xi_0 : \xi_1)$ are homogeneous coordinates of $\mathbb{P}^1$. The closure $\bar{N}'$ of N' in $N \times \mathbb{P}^1$ is an analytic subset. We set $M' = \pi^{-1}(M) \cap \bar{N}'$ with $\pi : N \times \mathbb{P}^1 \longrightarrow N$, the projection of $N \times \mathbb{P}^1$ onto its first factor. There exists the irreducible component M_1 of M' which is isomorphic to the graph of the meromorphic mapping of M into $\mathbb{P}^1$ defined by the meromorphic function y_{s+1} on M (see Example 2.4.1). The morphism $\pi_{M_1} : M_1 \longrightarrow M$ induced by the morphism π is a modification. Let N_1 be the irreducible component of $\bar{N}'$ containing M_1. In this situation, the natural morphism $\pi_{N_1} : N_1 \longrightarrow N$ is a generically finite morphism. Hence $a(N_1) = a(N)$. We let $h_1 : \bar{N}_1 \longrightarrow N_1$ be the normalization of N_1 and $\bar{M}_1 = h_1^{-1}(M_1)$. We restrict the meromorphic function ξ_0 / ξ_1 on N_1 to M_1 and obtain a meromorphic function x_{s+1} which is equal to $\pi^*_{M_1}(y_{s+1})$. Thus, the field $\mathbb{C}(\bar{N}_1)$ contains a subfield $L_1 = \mathbb{C}(x_1, x_2, \ldots, x_{s+1})$ isomorphic to F.

Applying to $\bar{N}_1$ and $\bar{M}_1$ the same process that we applied to $N = \bar{N}_0$ and $M = \bar{M}_0$, by induction, we obtain an infinite sequence

$$(\bar{N}_m, \bar{M}_m, h_m) \ , \quad m = 0, 1, 2, \ldots \ ,$$

satisfying the following conditions :

0) $\bar{h}_1 = \pi|_{N_1} \circ h_1$.

1) $\bar{M}_m$ is an irreducible subvariety of codimension one in a variety $\bar{N}_m$.

2) $\bar{h}_m : \bar{N}_{m+1} \longrightarrow \bar{N}_m$ is a generically finite morphism.

3) $\bar{N}_m$ is a normal variety.

4) $\bar{M}_{m+1} = \bar{h}_m^{-1}(\bar{M}_m)$.

5) $\bar{h}_m|\bar{M}_{m+1} : \bar{M}_{m+1} \longrightarrow \bar{M}_m$ is a modification.

6) Let $L(\bar{N}_m, \bar{M}_m)$ be the subring of the field $\mathbb{C}(\bar{N}_m)$ consisting of functions whose pole does not contain $\bar{M}_m$ and let $I(\bar{N}_m, \bar{M}_m)$ be the ideal in $L(\bar{N}_m, \bar{M}_m)$ consisting of functions which vanish on $\bar{M}_m$. Moreover, we set

$$F_m = L(\bar{N}_m, \bar{M}_m)/I(\bar{N}_m, \bar{M}_m)$$

and let

$$\tau_m : L(\bar{N}_m, \bar{M}_m) \longrightarrow F_m$$

be the natural homomorphism. Then $L(\bar{N}_{m+1}, \bar{M}_{m+1})$ contains a <u>subfield</u> L_{m+1} such that $\tau_{m+1}(L_{m+1}) = (\bar{h}_m|\bar{M}_{m+1})^*(F_m)$.

There is, therefore, a tower of fields

$$F_1 \subset F_2 \subset \cdots \subset F_m \subset F_{m+1} \subset \cdots \subset \mathbb{C}(M).$$

Since $\mathbb{C}(M)$ is finitely generated, for some positive integer m_0, we have

$$F_{m_0} = F_{m_0+1} = \cdots .$$

Clearly we have, $\tau_{m_0+1}(L_{m_0+1}) = F_{m_0+1}$. Hence $\bar{N}_{m_0+1}$ and $\bar{M}_{m_0+1}$ satisfy the assumption T). As we have

$$a(\bar{N}_{m_0+1}) = a(N), \quad a(\bar{M}_{m_0+1}) = a(M),$$

we have the desired result. Q.E.D.

§13. Complex manifolds of algebraic dimension zero

In this section we shall study structures of complex manifolds of algebraic dimension zero, using the results of the preceeding chapter.

In what follows, M is always assumed to be a compact complex manifold.

Lemma 13.1. If $a(M) = 0$, then the Albanese mapping $\alpha : M \longrightarrow A(M)$ is surjective. Hence $t(M) \leq \dim M$. Moreover $a(A(M)) = 0$.

Proof. Assume that the Albanese mapping α is not surjective. Then by Corollary 10.6, $\kappa(\alpha(M)) > 0$. As $\kappa(\alpha(M)) \leq a(\alpha(M))$, we have $a(M) \geq a(\alpha(M)) > 0$. This is a contradiction. Thus, α must be surjective and we have

$$0 = a(M) \geq a(A(M)) .$$ Q.E.D.

Lemma 13.2. If $a(M) = 0$, the Albanese dimension $t(M)$ of V is not equal to one.

Proof. If $t(M) = 1$, then $A(M)$ is an elliptic curve. Hence $a(A(M)) = 1$. This is a contradiction. Q.E.D.

To study the Albanese mapping more closely, we need the following lemma concerning divisors on a complex torus.

Lemma 13.3. If a complex torus T contains a prime divisor D, then $a(T) \geq 1$.

Proof. This is a corollary of Weil [2], Théorème 3, p.124. Here we shall provide another proof. Suppose that $\kappa(D) = 0$. Then, by Theorem 10.3, D is a translation of a complex subtorus T_1 of T. Then T/T_1 is an elliptic curve. Hence $a(T) \geq 1$.

Next suppose $\kappa(D) > 0$. Then by the proof of Theorem 10.9, there exist a complex subtorus T_1 and a subvariety D_1 of T/T_1 such that D is a T_1-bundle over D_1 and $\kappa(D_1) = \dim D_1 = \kappa(D)$. As D is a prime divisor, D_1 is a prime divisor on T/T_1. From the following lemma, we have $a(T) \geqq a(T/T_1) > 0$. Q.E.D.

Lemma 13.4. Let D be a prime divisor on a complex torus T such that $\kappa(D) = \dim D$. Then T is an abelian variety.

Proof. The smallest complex subtorus of T which contains D is T itself, since D is a divisor and $\kappa(D) > 0$. Hence D generates T. Hence, by Corollary 3.9 and Proposition 9.15, T is an abelian variety. Q.E.D.

Lemma 13.5. Assume that $a(M) = 0$. Let S be the smallest analytic subset of M such that the Albanese mapping $\alpha : M \longrightarrow A(M)$ is of maximal rank at any point of $M - S$. Then $\alpha(S)$ is of codimension at least two in $A(M)$.

Proof. By Corollary 1.7, $\alpha(S)$ is a nowhere dense analytic subset of $A(M)$. By Lemma 13.1 and Lemma 13.3, $A(M)$ does not contain divisors. Hence $\alpha(S)$ is of codimension at least two. Q.E.D.

Lemma 13.6. If $a(M) = 0$, then any fibre of the Albanese mapping is connected.

Proof. Let

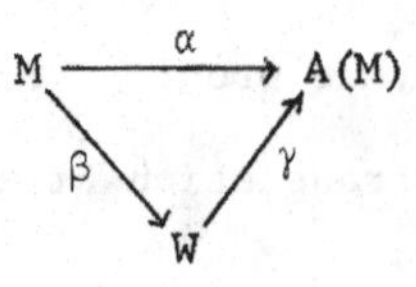

be the Stein factorization.

As $\alpha' = \alpha|_{M'} : M' = M - \alpha^{-1}(\alpha(S)) \longrightarrow A(M) - \alpha(S)$ is of maximal rank, $\gamma : W \longrightarrow A(M)$ is unramified on $A(M) - \alpha(S)$. As $\alpha(S)$ is of codimension at least two, there exists a finite unramified covering $\hat{\gamma} : \hat{W} \longrightarrow A(M)$ which is bimeromorphically equivalent to the finite morphism $\gamma : W \longrightarrow A(M)$. By Lemma 9.11, there exists a morphism $\hat{\beta} : M \longrightarrow \hat{W}$ which is bimeromorphically equivalent to the morphism $\beta : M \longrightarrow W$. From the universal property of the Albanese mapping (see Definition 9.6), we infer that $\hat{\gamma} : \hat{W} \longrightarrow A(M)$ is an isomorphism. Hence γ is an isomorphism on $A(M) - \gamma^{-1}(\alpha(S))$. This fact and Zariski's Main Theorem (Theorem 1.11) imply that γ is an isomorphism.

Q.E.D.

<u>Corollary 13.7</u>. If $a(M) = 0$, $t(M) = \dim M$, the Albanese mapping is a modification.

Next we shall study the case where $t(M) = \dim M - 1$.

<u>Theorem 13.8</u>. If $a(M) = 0$ and $t(M) = \dim M - 1$, then there exists an analytic subset T_1 of $A(M)$ of codimension at least two such that $\alpha|_{M'} : M' = M - \alpha^{-1}(T_1) \longrightarrow A(M) - T_1$ is an analytic fibre bundle whose fibre is a non-singular elliptic curve or $\mathbb{P}^1$.

<u>Proof</u>. By Lemma 13.5 there exists an analytic subset T_1 of $A(M)$ such that α is of maximal rank at any point of $M - \alpha^{-1}(T_1)$ and the codimension of T_1 in $A(M)$ is at least two. We set $T = A(M)$, $T' = T - T_1$ and $M' = \alpha^{-1}(M)$. By Lemma 13.6, for any point $x \in T'$, $M_x = \alpha^{-1}(x)$ is a non-singular curve of genus g. Suppose that $g \geq 1$. Let $\underline{M}_g$ be the moduli space of non-singular curves of genus g (see Baily [1] and Mumford [5], Chap.5 and Chap.7). $\underline{M}_g$ is a quasi projective variety of dimension $3g - 3$ for $g \geq 2$ and of

dimension one for $g = 1$. There exists a morphism $f : A(M) - T_1 \longrightarrow \underline{M}_g$ such that for $x \in T'$, $f(x)$ is a point of $\underline{M}_g$ corresponding to the isomorphism class of the curve M_x. Since $\underline{M}_g$ is quasi projective, if the image $f(T')$ is not a point, there exists a meromorphic function F on $\underline{M}_g$ such that $F \circ f$ is a non-constant meromorphic function on T'. Since T_1 is of codimension at least two in $A(M)$, $F \circ f$ can be extended to a non-constant meromorphic function on $A(M)$. This contradicts the assumption that $a(M) = 0$. Hence the image $f(T')$ must be a point. Hence M_x is isomorphic to a fixed curve C for any $x \in T'$. Therefore $\alpha' = \alpha|_{M'} : M' \longrightarrow T'$ is an analytic fibre bundle over T', whose fibre is the curve C.

Suppose that $g \geqq 2$. Since the automorphism group $\mathrm{Aut}(C)$ of the curve C is a finite group (see Corollary 14.3, below) and T_1 is a nowhere dense analytic subset, there exists an analytic fibre bundle $\hat{\alpha} : \hat{M} \longrightarrow A(M)$ over $A(M)$ whose fibre is the curve C such that $\hat{\alpha} : \hat{\alpha}^{-1}(T') \longrightarrow T'$ is analytically isomorphic to $\alpha' : M' \longrightarrow T'$.

Suppose that we can prove that $\alpha : M \longrightarrow A(M)$ is bimeromorphically equivalent to $\hat{\alpha} : \hat{M} \longrightarrow A(M)$. As $\mathrm{Aut}(C)$ is finite, there is an $\mathrm{Aut}(C)$-invariant non-constant meromorphic function G on C. Then G induces a non-constant meromorphic function on $\hat{M}$. This contradicts the assumption that $a(M) = 0$. Hence $g \leqq 1$.

Now we shall prove that, if $g \geqq 2$, $\alpha : M \longrightarrow A(M)$ is bimeromorphically equivalent to $\hat{\alpha} : \hat{M} \longrightarrow A(M)$.

For that purpose we use a result of hyperbolic analysis. For the definition of hyperbolic manifolds and detailed discussions, see Kobayashi [1].

<u>Theorem 13.9</u>. Let Y be a complex manifold and let Z be a complex submanifold of Y satisfying the following three conditions :

1) Z is hyperbolic ;

2) the closure of Z in Y is compact ;

3) given a point p of $\bar{Z} - Z$ and a neighbourhood U of p in Y, there exists a neighbourhood V of p in Y such that $\bar{V} \subset U$ and the distance between $Z \cap (Y - U)$ and $Z \cap V$ with respect to the hyperbolic distance d_Z is positive.

Let X be a complex manifold of dimension m and let A be a locally closed complex submanifold of dimension $\leqq m - 1$. Then every holomorphic mapping $f : X - A \longrightarrow Z$ can be extended to a holomorphic mapping from X into Y.

For the proof, see Kobayashi [1], Theorem 6.2, p.93-94.

We apply the above theorem to the following situation. For a point $x \in T_1$ we choose a neighbourhood U of x in $A(M)$ such that U is isomorphic to a polydisk D^t and $\hat{\alpha}^{-1}(U)$ is analytically trivial over U. We set $Z = \hat{\alpha}^{-1}(U) \cong D^t \times C$. As a universal covering of Z is a polydisk D^{t+1}, Z is a hyperbolic manifold (see Kobayashi [1], Theorem 4.7, p.58). The closure $\bar{Z}$ of Z in $\hat{M}$ is compact. Moreover it is easy to show that the above condition 3) is also satisfied. Let S_1 be the singular locus of the analytic set $\alpha^{-1}(T_1 \cap U)$. We set $X = \alpha^{-1}(U) - S_1$ and $A = \alpha^{-1}(T_1 \cap U) - S_1$. Then by the above theorem, the natural isomorphism $X - A$ into M can be extended to a holomorphic mapping of X into $\hat{M}$. Note that the image of $\alpha^{-1}(T_1 \cap U - S_1)$ is contained in $\hat{\alpha}^{-1}(T_1 \cap U - S_1)$. Again applying the above theorem to $X = \alpha^{-1}(U) - S_2$, $A = S_1 - S_2$, where S_2 is the singular locus of S_1, we obtain a holomorphic mapping of $\alpha^{-1}(U) - S_2$ into $\hat{M}$.

In this way we obtain a holomorphic mapping $\alpha^{-1}(U)$ into $\hat{M}$. Hence we have a holomorphic mapping of M into $\hat{M}$, which is the extension of the natural isomorphism between $\alpha^{-1}(A(M) - T_1)$ and $\hat{\alpha}^{-1}(A(V) - T_1)$. It follows that M and $\hat{M}$ are bimeromorphically equivalent.

Q.E.D.

In the above proof, we have used the important facts that the moduli space of non-singular curves of genus g exists and is quasi projective, hence it has many meromorphic functions. To generalize the above argument to the case where $t(V) = \dim V - 2$, we require the existance of the moduli space of algebraic surfaces of general type. For any surface S of general type there exists the minimal model $\hat{S}$ of S (see Theorem 20.3, below). $\hat{S}$ is characterized by the property that S does not contain exceptional curves of the first kind. We define $c_1^2 = K_{\hat{S}}^2$ and $p_a = p_g(S) - q(S) + 1$. These are birational invariants of the surface S which are invariant under deformations. The following theorem is due to H. Popp.

<u>Theorem 13.10</u>. The moduli space $\underline{M}_{c_1^2, p_a}$ of surface of general type with fixed c_1^2 and p_a exists as an algebraic space.

For the proof, see Popp [1], II, Chap. III.

Now we shall prove the following :

<u>Theorem 13.11</u>. If $a(M) = 0$ and $t(M) = \dim M - 2$, there exists an analytic subset T_1 of $A(M)$ of codimension at least two such that $\kappa(M_w) \leqq 0$ for any $w \in A(M) - T_1$. Moreover M_w is <u>not</u> a ruled surface of genus $g \geqq 2$.

<u>Proof</u>. We set $T = A(M)$. By Lemma 13.5 there exists an analytic

subset T_1 of T such that $\alpha : M \longrightarrow T$ is of maximal rank at any point of $M - \alpha^{-1}(T_1)$ and the codimension of T_1 in T is at least two. Thus, by Lemma 13.6, for any point $x \in T - T_1$, $M_x = \alpha^{-1}(x)$ is a non-singular surface. We set $T' = T - T_1$, $M' = \alpha^{-1}(T')$ and $\alpha' = \alpha_{|M'}$.

Suppose that $\kappa(M_x) = 2$ for a point $x \in T'$. Then $\kappa(M_y) = 2$ for any $y \in T'$ (see Theorem 20.11, below). As c_1^2 and p_a are invariant under deformations, there exists a morphism $f : T' \longrightarrow \underline{M}_{c_1^2, p_a}$ such that for $x \in T'$, $f(x)$ is a point corresponding to the isomorphism class of M_x. If the image $f(T')$ is not a point, there exists a non-constant meromorphic function F on $\underline{M}_{c_1^2, p_a}$ such that $F \circ f$ is a non-constant meromorphic function on T'. As T_1 is of codimension at least two, $F \circ f$ can be extended to a non-constant meromorphic function on T. Hence $a(T) \geqq 1$. This is a contradiction. Therefore, $f(T')$ must be a point. Then $\alpha' : M' \longrightarrow T'$ is an analytic fibre bundle whose fibre is a surface of general type S and whose structure group is Aut(S) (see Fischer and Grauert [1]). Since Aut(S) is a finite group (see Corollary 14.3, below) and T_1 is a nowhere dense analytic subset of T, the fibre bundle $\alpha' : M' \longrightarrow T'$ can be naturally extended to an analytic fibre bundle $\hat{\alpha} : \hat{M} \longrightarrow T$ over T whose fibre and structure group are S and Aut(S), respectively. Let G be an Aut(S)-invariant non-constant meromorphic function on S. The meromorphic function G induces a non-constant meromorphic function on $\hat{M}$. Hence $a(\hat{M}) \geqq 1$. We shall show that M is bimeromorphically equivalent to M. Since Aut(S) is a finite group, there exists a finite unramified covering $\tilde{T}$ of T such that the pull back $\tilde{\hat{\alpha}} : \tilde{\hat{M}} \longrightarrow \tilde{T}$ of $\hat{\alpha} : \hat{M} \longrightarrow T$ over $\tilde{T}$ is isomorphic to $p : \tilde{T} \times S \longrightarrow \tilde{T}$ of the projection of $\tilde{T} \times S$ onto the first factor $\tilde{T}$. Let $\tilde{\alpha} : \tilde{M} \longrightarrow \tilde{T}$ be the pull back of

$\alpha : M \longrightarrow T$ over $\tilde{T}$. M is bimeromorphically equivalent to $\hat{M}$ if and only if $\tilde{M}$ is bimeromorphically equivalent to $\tilde{\hat{M}}$. Hence, we can assume that $\hat{M}$ is a product $T \times S$. Let $p_1 : \hat{M} \longrightarrow T$ and $p_2 : \hat{M} \longrightarrow S$ be projections to the first and the second factors, respectively. Since the morphism $p_1 \circ \iota : M' \longrightarrow T$ is nothing other than α' where $\iota : M' \longrightarrow \hat{M}$ is a natural inclusion, the morphism $p_1 \circ \iota$ can be extended to a morphism $\pi_1 : M \longrightarrow T$. Therefore, it is enough to show that, for any point $x \in T$, there exists an open neighbourhood U of x in T such that the inclusion $\iota' : \alpha^{-1}(U) \cap M' \hookrightarrow \hat{\alpha}^{-1}(U)$ can be extended to a meromorphic mapping of $\alpha^{-1}(U)$ onto $\hat{\alpha}^{-1}(U)$. Since $\hat{\alpha}^{-1}(U) = U \times S$ and we can take U as a polydisk, it is easy to generalize the arguments in Kobayashi and Ochiai [1], Addendum p.143-p.148 (in this paper, they have considered the case where U is one point ; but since Theorem 4.4 in Kobayashi [1], Chap. II, p.28 is valid in the non-compact case, we can easily generalize their arguments) and we can prove that ι' can be extended to a meromorphic mapping. The details will be left to the reader.

Since M and $\hat{M}$ are bimeromorphically equivalent, we have $a(M) \geq 1$. This contradicts the assumption that $a(M) = 0$.

Next suppose that $\kappa(M_x) = 1$ for a point $x \in T'$. Then $\kappa(M_y) = 1$ for any $y \in T'$ (see Theorem 20.11, below). That is, M_y is an elliptic surface of general type (see Definition 20.6.2)). For a sufficiently large m (independent of $y \in T'$), the m-th canonical mapping

$$\Phi_{mK} : M_y \longrightarrow \mathbb{P}^N$$

is a morphism and its image $\Phi_{mK}(M_y)$ is a non-singular curve C_y of genus g (see 20.13, below). We write $\varphi_y = \Phi_{mK}$ and $\varphi_y : M_y \longrightarrow C_y$.

If $g = 0$, the elliptic surface $\varphi_y : M_y \longrightarrow C_y$ has at least three singular fibres. If $g = 1$, the elliptic surface $\varphi_y : V_y \longrightarrow C_y$ has at least one singular fibre. These facts are an easy consequence of the canonical bundle formula for an elliptic surface (see 20.13.1).

We set $\underline{F} = \alpha_*(\underline{O}(mK_V))$. Then by 2.10 there exists a meromorphic mapping $\psi : M \longrightarrow \mathbb{P}(\underline{F})$ such that for $x \in T'$, $\psi_x : M_x \longrightarrow \mathbb{P}(\underline{F})_x$ is nothing other than the m-th canonical mapping of M_x. Let $\tilde{M}$ be a non-singular model of the graph of the meromorphic mapping $\psi : M \longrightarrow \mathbb{P}(\underline{F})$ such that $\tilde{p}^{-1}(T') = \tilde{M}'$ is analytically isomorphic to M' and $\tilde{\psi}|_{\tilde{M}'} = \psi|_{M'}$ by this isomorphism where $\tilde{p} : \tilde{M} \longrightarrow T$ and $\tilde{\psi} : \tilde{M} \longrightarrow \mathbb{P}(\underline{F})$ are induced by the projection of the graph of ψ to the first and the second factor, respectively. As M and $\tilde{M}$ are bimeromorphically equivalent, we can assume that $\underline{\psi \text{ is a morphism}}$. Let W be the image of M under ψ and let $\pi : W \longrightarrow T$ be the morphism induced by the canonical morphism $\bar{p} : \mathbb{P}(\underline{F}) \longrightarrow T$. Then for any point $x \in T'$, $W_x = \pi^{-1}(x)$ is analytically isomorphic to the curve C_x.

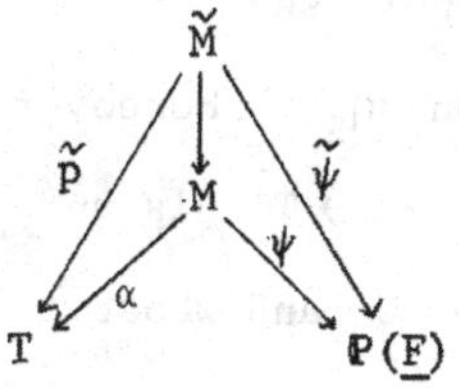

First suppose that the genus g of C_x is positive. As $a(W) = 0$, using the arguments in the proof of Theorem 13.8, we conclude that $g = 1$ and for any $x \in T'$, C_x is isomorphic to an elliptic curve C. Let S be one of the irreducible components of an analytic set $\psi(\{z \in V \mid d\psi(z) \text{ is not of maximal rank}\})$ such that for any point $x \in T'$, $\pi^{-1}(x) \cap S$ is not empty and is contained in the image of the union of singular fibres of the elliptic surface $\varphi_x : M_x \longrightarrow C_x$ under the morphism φ_x.

$$\begin{array}{ccc} S \subset W \subset \mathbb{P}(F) & & \\ \pi \downarrow \quad \swarrow \bar{p} & & \\ T & & \end{array}$$

As T does not contain any prime divisor, we conclude that the morphism $\pi_{|S} : S \longrightarrow T$ induces a finite unramified covering $\pi_{|S'} : S' \longrightarrow T'$ where $S' = \pi_{|S}^{-1}(T')$. As T_1 is of codimension at least two, there exists a finite unramified coveing $\hat{\pi}: \hat{S} \longrightarrow T$ of T that is the extension of $\pi_{|S'} : S' \longrightarrow T'$. Then S and $\hat{S}$ are bimeromorphically equivalent. We set $M^* = M \times_T \hat{S}$. M^* is a finite unramified covering of M and the morphism $\hat{\alpha} : M^* \longrightarrow \hat{S}$ induced by the projection to the second factor is nothing other than the Albanese mapping. We define W^* and S^* in a similar way as above. Then $\pi^*_{\ |S^*} : S^* \longrightarrow \hat{S}$ is a modification. Hence we can assume that $\pi_{|S} : S \longrightarrow T$ is a modification. This means that we have a meromorphic section $s : T \longrightarrow W$ defined by the inverse of the modification $\pi_{|S}$. Moreover on T', s is a morphism. Hence $\pi_{|W'} : W' = \pi^{-1}(T') \longrightarrow T'$ is an analytic fibre bundle whose fibre is an elliptic curve C and whose structure group G is a group of analytic automorphisms of the curve C which preserve the origin of C. As G is a finite group, the fibre bundle $\pi_{|W'} : W' \longrightarrow T'$ can be extended to a fibre bundle $\hat{\pi}: \hat{W} \longrightarrow T$ over T whose fibre is C and whose structure group is G. Moreover it is easy to show that W and $\hat{W}$ are bimeromorphically equivalent. Since G is a finite group, a G-invariant non-constant meromorphic function on C induces a non-constant meromorphic function on W. Hence we have $a(M) \geqq a(W) = a(\hat{W}) \geqq 1$. This contradicts the assumption that $a(M) = 0$.

If the genus g of C_x is zero, we use the similar argument as above we conclude that $a(M) \geqq a(W) \geqq 1$. In this case we use the above mentioned fact that the elliptic surface $\varphi_x : M_x \longrightarrow C_x$ has at least three singular fibres, for any point $x \in T'$. The details will be left to the reader.

Finally, suppose that M_x is a ruled surface of genus $g \geq 2$ for a point $x \in T'$. Then for any $y \in T'$, M_y is a ruled surface of genus g (see Theorem 20.11). Then a sheaf $\alpha'_*(\Omega^1_{V'})$ is a locally free sheaf of rank g. Hence for any point $y \in T'$, there exist an open neighbourhood U of y in T' and $\omega_1, \ldots, \omega_g$ of holomorphic 1-forms on $\alpha^{-1}(U)$ and integral (2q+1)-cycle $\gamma_1, \ldots, \gamma_{2g}$ of $\alpha^{-1}(U)$ such that $\omega_1, \ldots, \omega_g$ induce a basis $\{\omega_1(w), \ldots, \omega_g(w)\}$, of holomorphic 1-forms on M_w and $\{\gamma_1, \ldots, \gamma_{2g}\}$ induces a basis $\{\gamma_1(w), \ldots, \gamma_{2g}(w)\}$ of the free part of $H_1(M_w, \mathbb{Z})$ for any point $w \in U$. Moreover we can assume that we have a holomorphic section $s_U : U \longrightarrow \alpha^{-1}(U)$, since we can choose U sufficiently small. We set

$$\underline{a}_i(u) = \begin{pmatrix} \int_{\gamma_i(u)} \omega_1(u) \\ \int_{\gamma_i(u)} \omega_2(u) \\ \vdots \\ \int_{\gamma_i(u)} \omega_g(u) \end{pmatrix}, \quad u \in U.$$

Let G be a group of analytic automorphisms of $U \times \mathbb{C}^g$ consisting of automorphisms

$$g(n_1, n_2, \ldots, n_g) : U \times \mathbb{C}^g \longrightarrow U \times \mathbb{C}^g$$

$$\begin{matrix} \Psi & & \Psi \\ \left(u, \begin{pmatrix} \xi_1 \\ \vdots \\ \xi_g \end{pmatrix}\right) & \longmapsto & \left(u, \begin{pmatrix} \xi_1 \\ \vdots \\ \xi_g \end{pmatrix} + n_1\underline{a}_1(u) + n_2\underline{a}_2(u) + \cdots + n_g\underline{a}_g(u)\right) \end{matrix}$$

$$n_i \in \mathbb{Z}, \quad i = 1, 2, 3, 4.$$

The group G acts on $U \times \mathbb{C}^g$ properly discontinuously and freely. The quotient manifold $A_U = U \times \mathbb{C}^g/G$ has the structure of a fibre space of abelian varieties over U induced by the projection of $U \times \mathbb{C}^g$ onto U. Now we define a morphism

$$\begin{array}{cccl} f_u : \alpha^{-1}(u) & \longrightarrow & A_u & \\ \cup & & \cup & \\ z & \longmapsto & [\alpha(z), {}^t(\displaystyle\int_{s_U(\alpha(z))}^{z} \omega_1(\alpha(z)) , & \\ & & \displaystyle\int_{s_U(\alpha(z))}^{z} \omega_2(\alpha(z)), \ldots, \int_{s_U(\alpha(z))}^{z} \omega_g(\alpha(z)))] . & \end{array}$$

The image $C_U = f_U(\alpha^{-1}(U))$ is non-singular and the morphism $\pi_U : C_U \longrightarrow U$ induced by the natural morphism of A_U onto U is of maximal rank at any point C_U. For any $u \in U$, $\pi_U^{-1}(u)$ is a non-singular curve of genus g. In this way we have an open covering $\{U_i\}$ of T', fibre spaces $\pi_{U_i} : C_{U_i} \longrightarrow U_i$ of non-singular curves of genus g and surjective morphisms $f_{U_i} : \alpha^{-1}(U_i) \longrightarrow C_{U_i}$. As a consequence of our construction, we can patch together the C_{U_i}'s and obtain a fibre space $\pi' : \mathcal{C}' \longrightarrow T'$ of non-singular curves of genus g and a morphism $f' : V' \longrightarrow \mathcal{C}'$. Using the arguments of the proof of Theorem 13.8, we conclude that $\pi' : \mathcal{C}' \longrightarrow T'$ is an analytic fibre bundle over T' whose fibre is a non-singular curve C of genus g and whose structure group is $\mathrm{Aut}(C)$. As $\mathrm{Aut}(C)$ is finite, we can extend the fibre bundle $\pi' : \mathcal{C}' \longrightarrow T'$ to an analytic fibre bundle $\pi : \mathcal{C} \longrightarrow T$ over T whose fibre and structure group are C and $\mathrm{Aut}(C)$, respectively. Then we have seen that $a(\mathcal{C}) \geqq 1$. On the other hand, by Theorem 13.9, the morphism $f' : M' \longrightarrow \mathcal{C}'$ can be extended to a morphism $f : M \longrightarrow \mathcal{C}$. Hence we have $a(M) \geqq a(\mathcal{C}) \geqq 1$.

This is a contradiction. Q.E.D.

Remark 13.12. The classification of analytic surfaces shows that a general fibre of the Albanese mapping $\alpha : M \longrightarrow A(M)$ in Theorem 13.11 is one of the following surfaces :

(1) K3 surface. (2) complex torus. (3) hyperelliptic surface.
(4) Enriques surface. (5) elliptic surface with a trivial canonical bundle. (6) surface of class VII. (7) rational surface.
(8) ruled surface of genus one.

The author does not know whether hyperelliptic surfaces and Enriques surfaces appear as fibres.

Examples 13.13. In what follows, T is always assumed be a two dimensional complex torus with a period matrix $\Omega = (a_1, a_2, a_3, a_4)$ such that $a(T) = 0$. For example, $\Omega = \begin{pmatrix} 1 & 0 & \sqrt{-2} & \sqrt{-5} \\ 0 & 1 & \sqrt{-3} & \sqrt{-7} \end{pmatrix}$.

We construct a four-dimensional complex manifold M such that $a(M) = 0$, $t(M) = 2$, $A(M) \cong T$ and the Albanese mapping has the structure of an analytic fibre bundle over T whose fibre is a surface of type (1), (2), (5), (6), (7), (8) in Remark 18.12.

1) Let S be a K3 surface of algebraic dimension zero. Then $V = T \times S$ satisfies the preceeding conditions.

2) Let A be an abelian surface. We choose four points $b_1, b_2, b_3, b_4 \in A$ in general position. Let G be a group of analytic automorphisms of $\mathbb{C} \times A$ generated by automorphisms

$$g_i : \mathbb{C}^2 \times A \longrightarrow \mathbb{C}^2 \times A, \quad i = 1, 2, 3, 4.$$
$$(\zeta, z) \longmapsto (\zeta + a_i, -z + b_i).$$

The group G acts on $\mathbb{C}^2 \times A$ properly discontinuously and freely. The quotient manifold $M = \mathbb{C}^2 \times A/G$ has the structure of a fibre bundle over $T = \mathbb{C}^2/\Omega$ induced by the natural projection $p_1 : \mathbb{C}^2 \times A \longrightarrow \mathbb{C}^2$. The fibre of this bundle is an abelian variety A. By our construction $H^0(V, \Omega^1_V)$ is spanned by dz_1 and dz_2. Hence $A(M) \cong T$. Moreover it is easy to see that $\mathbb{C}(M) = \mathbb{C}$.

3) An elliptic surface with a trivial canonical bundle is represented by the quotient manifold $\mathbb{C}^2/G$ where G is a group of analytic automorphisms of $\mathbb{C}^2$ generated by automorphisms

$$g_j : \begin{array}{l} z_1 \longmapsto z_1 + \alpha_j \\ z_2 \longmapsto z_2 + \bar{\alpha}_j z_1 + \beta_j, \quad j = 1, 2, 3, 4, \end{array}$$

such that

$$\alpha_1 = \alpha_2 = 0,$$
$$\bar{\alpha}_3 \alpha_4 - \bar{\alpha}_4 \alpha_3 = m\beta_2 \neq 0,$$

where m is a positive integer, and $\{\alpha_3, \alpha_4\}$, $\{\beta_1, \beta_2\}$ are fundamental periods of two elliptic curves (see Kodaira [3], I, p.786~ p.788).

Here, we set

$$\alpha_3 = 1, \quad \alpha_4 = b\sqrt{-1}, \quad \beta_1 = \frac{2b^2}{1+b^2} - \frac{2b^3}{1+b^2}\sqrt{-1}$$
$$\beta_2 = 2b\sqrt{-1}, \quad \beta_3 = \beta_4 = 0,$$

where b is a transcendental number. We construct a surface $S = \mathbb{C}^2/G$ with these α_i's and β_i's.

Let $\mathfrak{G}$ be a group of analytic automorphisms of $\mathbb{C}^2 \times S$ generated by automorphisms

$$g_1 : (\zeta, [z_1, z_2]) \longrightarrow (\zeta + a_1, [z_1 + \alpha, z_2])$$
$$g_j : (\zeta, [z_1, z_2]) \longrightarrow (\zeta + a_j, [-z_1, z_2]), \ j=2,3,4,$$

where

$$\alpha = \frac{2b^2}{1 + b^2} + \frac{2b}{1 + b^2}\sqrt{-1}$$

and $[z_1, z_2]$ is a point of S corresponding to a point $(z_1, z_2) \in \mathbb{C}^2$ Let $M = \mathbb{C}^2 \times S/\mathcal{G}$ be the quotient manifold. The projection $\mathbb{C}^2 \times S \longrightarrow \mathbb{C}^2$ induces a surjective morphism $f : M \longrightarrow T = \mathbb{C}^2/\Omega$. By the morphism f, M is an analytic fibre bundle over T whose fibre is S. By our construction $H^0(V, \Omega_V^1)$ is spanned by $d\zeta_1$ and $d\zeta_2$. Hence $A(M) \cong T$. Let E be an elliptic curve with fundamental periods $\{\alpha_1, \alpha_2\}$. Then there exists a surjective morphism

$$g : S \longrightarrow E$$
$$[z_1, z_2] \longmapsto [z_1]$$

such that, via g, we have $\mathbb{C}(S) \cong \mathbb{C}(E)$. Let H be a group of analytic automorphisms of the elliptic curve E generated by automorphisms

$$h_1 : [z_1] \longmapsto [z_1 + \alpha]$$
$$h_2 : [z_2] \longmapsto [-z_1] .$$

Then we have

$$\mathbb{C}(M) \cong \mathbb{C}(E)^H = \mathbb{C} .$$

Hence M has the desired properties.

In the similar way we can construct an analytic fibre bundle M over T whose fibre is a surface of class VII, rational surface or ruled surface of genus one. We leave the construction to the reader.

Chapter VI

Addition formula for Kodaira dimensions of analytic fibre bundles

The main purpose of the present chapter is to prove an addition formula for the Kodaira dimension of an analytic fibre bundles whose fibre and structure group are a Moishezon manifold F and the analytic automorphism group $Aut(F)$ of F, respectively (see Theorem 15. 1, below). For that purpose, in §14, we shall study the action of a bimeromorphic transformation group $Bim(V)$ of a complex manifold V on a vector space $H^0(V, \underline{O}(mK_V))$. From this action, we have a pluri-canonical representation

$$\rho_m : Bim(V) \longrightarrow GL(H^0(V, \underline{O}(mK_V))).$$

We shall prove that, if V is a Moishezon manifold, the image $\rho_m(Bim(V))$ is a finit group, but if V is not a Moishezon manifold $\rho_m(Bim(V))$ is not necessarily a finite group (see Theorem 14.10 and Remark 14.6). This is the reason why the fibres need to be Moishezon manifolds in the addition formula for Kodaira dimensions of fibre bundles.

In §15, the addition formula will be proved. This gives an affirmative answer to Conjecture C_n in §11 in our situation. To prove the addition formula, we only need the fact that for any element $g \in Bim(V)$, $\rho_m(g)$ is of finite order. But this may not be the case if V is not a Moishezon manifold (see Remark 14.6).

This chapter is taken from Nakamura and Ueno [1].

§14. Pluricanonical representations of bimeromorphic transformation groups

Let V be a complex variety. By $\mathrm{Bim}(V)$ (resp. $\mathrm{Aut}(V)$), we mean the group of all bimeromorphic mappings of V onto itself (resp. a group of all automorphisms of V). $\mathrm{Bim}(V)$ (resp. $\mathrm{Aut}(V)$) is called the bimeromorphic transformation group (the analytic automorphism group) of V. If V is an algebraic variety, $\mathrm{Bim}(V)$ will be written as $\mathrm{Bir}(V)$ and is called the birational transformation group of V.

Under the compact open topology, $\mathrm{Aut}(V)$ becomes a complex Lie group (see Kaup [1], Douady [1], p.92). Let $\mathrm{Aut}^0(V)$ be the identity component of $\mathrm{Aut}(V)$. If V is an algebraic variety, then $\mathrm{Aut}^0(V)$ is an algebraic group (Grothendieck [1], Matsumura and Oort [1]). $\mathrm{Aut}(V)/\mathrm{Aut}^0(V)$ may not be a finite group for a variety V.

$\mathrm{Bim}(V)$ is in general large.

By Lemma 6.3, $\mathrm{Bim}(V)$ operates on $H^0(V, \underline{O}(mK_V))$ for any positive integer m. Hence we have a group representation

$$\rho_m : \mathrm{Bim}(V) \longrightarrow GL(H^0(V, \underline{O}(mK_V))).$$

We call this representation a pluricanonical representation of $\mathrm{Bim}(V)$. The main purpose of this section is to study pluricanonical representations and to prove that $\rho_m(\mathrm{Bim}(V))$ is a finite group for a Moishezon manifold (see Theorem 14.10). For that purpose we need the following theorem which is a part of a theorem due to Matsumura [1].

Theorem 14.1. Let V be an algebraic variety. If $\mathrm{Bir}(V)$ contains a linear algebraic group G, then V is a ruled variety. That is, V is birationally equivalent to a product $\mathbb{P}^1 \times W$ where W is an algebraic variety. Hence a fortiori, $\kappa(V) = -\infty$.

Proof. A linear algebraic group G contains a one-dimensional linear algebraic group H. As $H \subset \mathrm{Bir}(V)$, H operates on the function field $\mathbb{C}(V)$ of V. Rosenlicht has shown that the invariant field $\mathbb{C}(V)^H$ of $\mathbb{C}(V)$ by H is an algebraic function field and that $\mathbb{C}(V)$ is isomorphic to $\mathbb{C}(V)^H(t)$ where t is transcendental over $\mathbb{C}(V)^H$ (see Rosenlicht [1], Theorem 10, p.426). This implies that V is birationally equivalent to $\mathbb{P}^1 \times W$ where $\mathbb{C}(W)$ is isomorphic to $\mathbb{C}(V)^H$.

Q.E.D.

Remark 14.2. Suppose that, for a compact complex manifold V, $\mathrm{Bim}(V)$ contains a linear algebraic group. It is not known whether $\kappa(V) = -\infty$ or not.

Corollary 14.3. Let V be a complex manifold of hyperbolic type (i.e., $\kappa(V) = \dim V$). Then $\mathrm{Bim}(V)$ is a finite group.

Proof. For a sufficiently large positive integer m, the m-th canonical mapping

$$\Phi_{mK} : V \longrightarrow \mathbb{P}^N$$

of V is bimeromorphic. We set $\hat{V} = \Phi_{mK}(V)$. As g operates on $H^0(V, \underline{O}(mK_V))$, g induces a projective transformation $\tilde{g}$ of $\mathbb{P}^N$ which leaves $\hat{V}$ invariant. Hence $\tilde{g}$ induces an automorphism $\hat{g}$ of $\hat{V}$. Let $\mathrm{Lin}(\hat{V})$ be a subgroup of $\mathrm{Aut}(\hat{V})$ consisting of the automorphisms of $\hat{V}$ induced by projective transformations of the ambient space $\mathbb{P}^N$ which leaves $\hat{V}$ invariant. The group $\mathrm{Lin}(\hat{V})$ is obviously a linear algebraic group. As $\kappa(\hat{V}) = \kappa(V) = \dim V$, by Theorem 14.1, $\mathrm{Lin}(\hat{V})$ must be a discrete group. Therefore, $\mathrm{Lin}(\hat{V})$ is a finite group (note that an algebraic group has a finite number of connected components). On the other hand as Φ_{mK} is bimeromorphic, the group homomorphism

$$\begin{array}{ccc} \mathrm{Bim}(V) & \longrightarrow & \mathrm{Lin}(\hat{V}) \\ \cup & & \cup \\ g & \longmapsto & \hat{g} \end{array}$$

is injective. Hence $\mathrm{Bim}(V)$ is a finite group. Q.E.D.

Let $g : W \longrightarrow V$ be a meromorphic mapping of a complex manifold W into a complex manifold V of the same dimension n. Let W^* be a non-singular model of the graph of g with canonical projections f and h,

$$\begin{array}{ccc} & W^* & \\ {}^{f}\swarrow & & \searrow^{h} \\ W & \xrightarrow{\ g\ } & V \end{array} .$$

We consider a homomorphism $f_*^{2n-k} : H_{2n-k}(W^*, \mathbb{Z})_0 \longrightarrow H_{2n-k}(W, \mathbb{Z})_0$ of the free parts of homology groups induced by f. We define a homomorphism $f_{k*} : H^k(W^*, \mathbb{Z})_0 \longrightarrow H^k(W, \mathbb{Z})_0$ of the free parts of cohomology groups by

$$\begin{array}{ccc} H^k(W^*, \mathbb{Z})_0 & \longrightarrow & H^k(W, \mathbb{Z})_0 \\ \cup & & \cup \\ x & \longmapsto & (f_*^{2n-k}(x^d))^d \end{array} ,$$

where "d" denotes the Poincaré dual operator. Thus, setting $g_k^* = f_{k*} \circ h_k^*$, where $h_k^* : H^k(V, \mathbb{Z})_0 \longrightarrow H^k(W^*, \mathbb{Z})_0$ is a homomorphism induced by h, we have a homomorphism $g_k^* : H^k(V^*, \mathbb{Z})_0 \longrightarrow H^k(W, \mathbb{Z})_0$ of the free parts of cohomology groups. It is easy to check that the definition of g_k^* is independent of the choice of W^*. g_k^* can be extended to a homomorphism $g_k^* : H^k(V, \mathbb{C}) \longrightarrow H^k(W, \mathbb{C})$.

As a holomorphic n-form on an n-dimensional complex manifold is d-closed, by de Rham's theorem we can regard $H^0(V, \underline{O}(K_V))$ and $H^0(W, \underline{O}(K_W))$ as subspaces of $H^0(V, \mathbb{C})$ and $H^0(W, \mathbb{C})$, respectively. Then for any element $\varphi \in H^0(V, \underline{O}(K_V))$, we have

$$g^*(\varphi) = g_n^*(\varphi) .$$

Proposition 14.4. Let g be a bimeromorphic mapping of an n-dimensional complex manifold V onto itself. If we have

$$g^*(\varphi) = \alpha\varphi , \quad \alpha \in \mathbf{C},$$

for some non zero element $\varphi \in H^0(V, \underline{O}(mK_V))$, then α is an algebraic integer. Moreover, the degree $[\mathbf{Q}(\alpha) : \mathbf{Q}]$ of the algebraic extension $\mathbf{Q}(\alpha)$ over $\mathbf{Q}$ is bounded above by some constant $N(\varphi)$, which depends on φ but does not depend on the bimeromorphic mapping g.

Proof. First suppose that $m = 1$. φ is a holomorphic n-form. Since we have

$$g^*(\varphi) = g_n^*(\varphi),$$

α is an eigenvalue of the automorphism g_n^* of $H^n(V, \mathbf{Z})_0$. It follows that α is an algebraic integer. The degree of the minimal equation of α with coefficients in $\mathbf{Q}$ is bounded above by the n-th Betti number $b_n(V)$ of V.

Next suppose that $m \geqq 2$. Let $\{V_i\}_{i\in I}$ be a sufficiently fine finite open covering of V such that $(z_i^1, z_i^2, \ldots, z_i^n)$ is a system of local coordinates in V_i. In terms of these local coordinates, φ can be expressed in the form

$$\varphi_i(z_i^1, \ldots, z_i^n)(dz_i^1 \wedge \ldots \wedge dz_i^n)^m ,$$

where φ_i is holomorphic on V_i. Let $\mathbb{K}$ be a non-compact complex manifold which is the total space of the canonical line bundle K_V of V. $\mathbb{K}$ is covered by coordinate neighbourhoods U_i with a system of coordinates $(z_i^1, z_i^2, \ldots, z_i^n, w_i)$ such that U_i is analytically isomorphic to $V_i \times \mathbf{C}$. Let V' be the subvariety of $\mathbb{K}$ defined by equations

$$(w_i)^m = \varphi_i(z_i^1, \ldots, z_i^n) ,$$

for any $i \in I$. V' may not be connected. The holomorphic n-forms $w_i dz_i^1 \wedge \cdots \wedge dz_i^n$ on U_i, $i \in I$, define a global holomorphic n-form ψ on $\mathbb{K}$. Moreover, a bimeromorphic mapping g induces a bimeromorphic mapping $g_{\mathbb{K}}$ of $\mathbb{K}$ onto itself. In fact, if $g(V_i) \cap V_j \neq \phi$, $g_{\mathbb{K}}|_{U_i'}$: $U_i' = g^{-1}(g(V_i) \cap V_j) \times \mathbb{C} \longrightarrow V_j \times \mathbb{C}$ can be expressed in terms of the above local coordinates in the form

$$(z_i^1, \ldots, z_i^n, w_i) \longrightarrow (g^1(z_i), \ldots, g^n(z_i), \det\left(\frac{\partial(g^1(z_i), \ldots, g^n(z_i))}{\partial(z_i^1, \ldots, z_i^n)}\right)^{-1} w_i)$$

Let m_β be an analytic automorphisms of $\mathbb{K}$ defined by

$$m_\beta : (z_i^1, \ldots, z_i^n, w_i) \longmapsto (z_i^1, \ldots, z_i^n, \beta w_i), \quad i \in I ,$$

where β is one of the m-th roots of α. Since $g^*(\varphi) = \alpha \varphi$, the bimeromorphic mapping $m_\beta . g_{\mathbb{K}}$ induces a bimeromorphic mapping of V' onto V'.

By a suitable sequence of monoidal transformations of the manifold $\mathbb{K}$ with non-singular centers, we obtain a manifold $\tilde{\mathbb{K}}$ and the strict transform W of V', which is a non-singular model of the variety V' (see Theorem 2.12). Then the bimeromorphic mapping $m_\beta . g_{\mathbb{K}}$ of $\mathbb{K}$ can be lifted to a bimeromorphic mapping $\tilde{h}$ of $\tilde{\mathbb{K}}$ which induces a bimeromorphic mapping h of W onto itself.

Let $f_1 : W \longrightarrow V'$ be the surjective morphism induced from the above monoidal transformations. Let $f_2 : V' \longrightarrow V$ be a finite surjective morphism defined by

$$f_2 : (z_i^1, \ldots, z_i^n, w_i) \longrightarrow (z_i^1, \ldots, z_i^n) .$$

We set $f = f_2 \circ f_1$.

The holomorphic n-form ψ can be lifted to a holomorphic n-form $\tilde{\psi}$ on $\tilde{\mathbb{K}}$, which induces a holomorphic n-form ω on W.

From the arguments above, it follows without difficulty that

$$(\omega)^m = f^*(\varphi) \quad .$$

Moreover, since $(m_\beta \cdot g_{IK})^*(\Psi) = \beta\Psi$, we have

$$h^*(\omega) = \beta\,\omega \quad .$$

From the first part of the proof, we infer that β is an algebraic integer and $[\mathbb{Q}(\beta) \; ; \; \mathbb{Q}] \leqq b_n(W)$. This implies that $\alpha = \beta^m$ is an algebraic integer and $[\mathbb{Q}(\alpha) \; ; \; \mathbb{Q}] \leqq b_n(W)$. The number $b_n(W)$ depends only on φ but does not depend on g. Q.E.D.

<u>Proposition 14.5</u>. Let V, g, φ and α be the same as those of Proposition 14.4. Then we have $|\alpha| = 1$. Moreover, when V is a Moishezon manifold, α is a root of unity.

<u>Proof</u>. We use the same notations as above. By $(\varphi \wedge \bar{\varphi})^{\frac{1}{m}}$ we denote a differentiable 2n-form on V which has the form

$$(\sqrt{-1})^{-n^2} |\varphi_i(z_i)|^{\frac{2}{m}} dz_i^1 \wedge \cdots \wedge dz_i^n \wedge d\bar{z}_i^1 \wedge \cdots \wedge d\bar{z}_i^n ,$$

over V_i. We set

$$\| \varphi \| = \left(\int_V (\varphi \wedge \varphi)^{\frac{1}{m}} \right)^{\frac{1}{2}} \quad .$$

Then we have

$$0 < \|\varphi\|^2 = \int_V (\varphi \wedge \bar{\varphi})^{\frac{1}{m}} = \int_V (g^*\varphi \wedge \overline{g^*\varphi})^{\frac{1}{m}} = \|g^*\varphi\|^2 = |\alpha|^{\frac{2}{m}} \|\varphi\|^2 .$$

Hence $|\alpha| = 1$.

Next we shall prove the latter half of the proposition. Since any Moishezon manifold is bimeromorphically equivalent to a projective manifold, we can assume that V is projective. We fix an imbedding of V into $\mathbb{P}^N$ for some N and let $I(V)$ be the defining ideal of V. For an automorphism σ of the complex number field and a polynomial $f(z) = f(z_0, \ldots, z_N)$, we define $f^\sigma(z) = (f(z_0^{\sigma^{-1}}, \ldots, z_N^{\sigma^{-1}}))^\sigma$ and also define $I(V)^\sigma = \{f^\sigma \,;\, f \in I(V)\}$. The ideal $I(V)^\sigma$ defines a projective manifold V^σ. In that case, the meromorphic mapping g^σ of V^σ is defined to be $g^\sigma(z) = (g(z^{\sigma^{-1}}))^\sigma$, symbolically. Similarly, for an element $\varphi \in H^0(V, \underline{O}(mK_V))$, we define $\varphi^\sigma \in H^0(V^\sigma, \underline{O}(mK_{V^\sigma}))$. Then we have

$$\begin{aligned}(g^\sigma)^*(\varphi^\sigma(z)) &= \varphi^\sigma(g^\sigma(z)) = (\varphi((g^\sigma(z))^{\sigma^{-1}}))^\sigma \\ &= (\varphi((g(z^{\sigma^{-1}})^\sigma)^{\sigma^{-1}}))^\sigma = (\varphi(g(z^{\sigma^{-1}})))^\sigma \\ &= (g^*(\varphi))^\sigma(z) .\end{aligned}$$

Hence, if $g^*(\varphi) = \alpha\varphi$, then we have $g^{\sigma *}(\varphi^\sigma) = \alpha^\sigma\varphi^\sigma$. The above arguments implies that $|\alpha^\sigma| = 1$. Hence α is a root of unity.

Q.E.D.

Remark 14.6. Proposition 14.5 does not hold for an arbitrary complex manifold. S. Iitaka has constructed the following example.

Let a,b,c be three roots of the equation

$$x^3 + 3x + 1 = 0 ,$$

such that a is real. Let $\alpha_1, \alpha_2, \beta_1, \beta_2, \gamma_1, \gamma_2$ be six roots of the equation

$$z^6 + 3z^2 + 1 = 0 ,$$

such that

$$\alpha_i^2 = a,\ \beta_i^2 = b,\ \gamma_i^2 = c,\quad i = 1,2.$$

We set

$$\Omega = \begin{pmatrix} 1 & \alpha_1 & \alpha_1^2 & \alpha_1^3 & \alpha_1^4 & \alpha_1^5 \\ 1 & \beta_1 & \beta_1^2 & \beta_1^3 & \beta_1^4 & \beta_1^5 \\ 1 & \beta_2 & \beta_2^2 & \beta_2^3 & \beta_2^4 & \beta_2^5 \end{pmatrix} .$$

There exists a three-dimensional complex torus T with a period matrix Ω. Left multiplication of a matrix $\begin{pmatrix} \alpha_1 & 0 & 0 \\ 0 & \beta_1 & 0 \\ 0 & 0 & \beta_2 \end{pmatrix}$ to $\mathbb{C}^3$ induces an analytic automorphism g of the complex torus T. Then we have

$$g^*(dz_1 \wedge dz_2 \wedge dz_3) = \alpha\, dz_1 \wedge dz_2 \wedge dz_3 \quad ,$$

where z_1, z_2, z_3 are global coordinates of T and

$$\alpha = \alpha_1 \beta_1 \beta_2 = -\alpha_1 b \quad .$$

On the other hand, the Galois group of $L = \mathbb{Q}(a,b,c)$ over $\mathbb{Q}$ is a symmetric group S_3. Hence there exists an automorphism σ of the field $\mathbb{Q}(\alpha_1, \alpha_2, \beta_1, \beta_2, \gamma_1, \gamma_2)$ such that $\alpha^\sigma = \beta_1 \gamma_1 \gamma_2 = -\beta_1 c$. Since

$$|a| < 1, \quad |c| = 1/\sqrt{|a|} > 1, \quad |\beta_1| = \sqrt{|c|} \ ,$$

we have

$$|\alpha^\sigma| = \sqrt{|c|^3} > 1.$$

Hence α is not a root of unity.

<u>Proposition 14.7</u>. Let V be a complex manifold and let ρ_m : $\mathrm{Bim}(V) \longrightarrow GL(H^0(V, \underline{O}(mK_V))$ be a pluricanonical representation. Then for any element $g \in \mathrm{Bim}(V)$, $\rho_m(g)$ is semi-simple.

<u>Proof</u>. If $\rho_m(g)$ is not semi-simple there exists two linearly independent elements φ_1, φ_2 of $H^0(V, \underline{O}(mK_V))$ such that

$$g^*(\varphi_1) = \alpha \varphi_1 + \varphi_2$$

$$g^*(\varphi_2) = \alpha \varphi_2 \ ,$$

where α is an algebraic integer by Proposition 14.4 and $|\alpha| = 1$ by Proposition 14.5. We have

$$(g^\ell)^*(\varphi_1) = \alpha^\ell \varphi_1 + \ell \alpha^{\ell-1} \varphi_2 \ .$$

Since g is a bimeromorphic mapping of V onto itself, we have

$$\| (g^\ell)^*(\varphi_1) \| = \| \varphi_1 \| \ .$$

On the other hand we have

$$\|(g^{\ell})^{*}(\varphi_1)\|^2 = (\sqrt{-1})^{-n^2}\int_V |\alpha^{\ell}\varphi_{1,i} + \ell\alpha^{\ell-1}\varphi_{2,i}|^{\frac{2}{m}}\, dz_i^1\wedge\cdots\wedge dz_i^n\wedge d\bar z_i^1\wedge\cdots\wedge d\bar z_i^n$$

$$= (\sqrt{-1})^{-n^2}\ell^{\frac{2}{m}}\int_V \left|\frac{\varphi_{1,i}}{\ell} + \frac{\varphi_{2,i}}{\alpha}\right|^{\frac{2}{m}} dz_i^1\wedge\cdots\wedge dz_i^n\wedge d\bar z_i^1\wedge\cdots\wedge d\bar z_i^n.$$

It is easy to see that there exists a positive number A such that

$$(\sqrt{-1})^{-n^2}\int_V \left|\frac{\varphi_{1,i}}{\ell} + \frac{\varphi_{2,i}}{\alpha}\right|^{\frac{2}{m}} dz_i^1\wedge\cdots\wedge dz_i^n\wedge d\bar z_i^1\wedge\cdots\wedge d\bar z_i^n \geqq A$$

for any positive integer ℓ. Hence

$$\lim_{\ell\to+\infty}\|(g^{\ell})^{*}(\varphi_1)\|^2 = +\infty .$$

This contradicts the fact $\|\varphi_1\| = \|(g^{\ell})^{*}(\varphi_1)\|$. Q.E.D.

Corollary 14.8. For a complex manifold V, let $\mathrm{Aut}^0(V)$ be the identity component of $\mathrm{Aut}(V)$. Then $\rho_m(\mathrm{Aut}^0(V))$ is the identity matrix.

Proof. It is enough to show that $\dim \rho_m(\mathrm{Aut}^0(V)) = 0$. Assume the contrary. We set $G = \rho_m(\mathrm{Aut}^0(V))$. G is a connected complex Lie group. Let $\mathfrak{g}$ be the complex Lie algebra of G. For any non zero element $X \in \mathfrak{g}$, let $\exp(tX)$ be a complex one parameter subgroup of G. As any element of G is semi-simple, there exists $N \times N$ matrix M ($N = \dim H^0(V, \underline{O}(mK_V))$) such that

$$M\exp(tX)M^{-1} = \begin{pmatrix} a_1(t) & & & & 0 \\ & a_2(t) & & & \\ & & \cdot & & \\ & & & \cdot & \\ 0 & & & \cdot & a_N(t) \end{pmatrix}$$

where $a_i(t)$ is holomorphic on $\mathbb{C}$. On the other hand, by Proposition 14.5 we have $|a_i(t)| = 1$ for any t. Hence $a_i(t)$ must be a constant. Therefore $a_i(t) = a_i(0) = 1$. As G is generated by one parameter subgroups, this contradicts the assumption that

$\dim \rho_m(\mathrm{Aut}^0(V)) \geqq 1$. Q.E.D.

To obtain the main theorem in this section, we need a certain classical result on subgroup of general linear group. A subgroup G of $GL(n, \mathbb{C})$ is called a <u>periodic subgroup</u> if every element of G is of finite order.

<u>Theorem 14.9</u>. (Burnside) Let G be a periodic subgroup of $GL(n, \mathbb{C})$. If the order of any element g of G is uniformly bounded, G is a finite group.

For the proof, see Curtis and Reiner [1], §36.1, p.251.

Finally we arrive at the main theorem in this section.

<u>Theorem 14.10</u>. Let V be a Moishezon manifold. Then $\rho_m(\mathrm{Bim}(V))$ is a finite group.

<u>Proof</u>. By Theorem 14.9, it is enough to show that for any element $g \in \mathrm{Bim}(V)$, the order of $\rho_m(g)$ is uniformly bounded. Since $\rho_m(g)$ is diagonizable by Proposition 14.7 and its eigenvalue is root of unity by Proposition 14.4, it is enough to show that the constant $N(\varphi)$ in Proposition 14.4 is uniformly bounded for any element $\varphi \in H^0(V, \underline{O}(mK_V))$. For that purpose we shall prove that a non-singular model W of a branched covering V' of V constructed by φ in the proof of Proposition 14.4 can be chosen in such a way that the n-th Betti number $b_n(W)$ is uniformly bounded for any $\varphi \in H^0(V, \underline{O}(mK_V))$. From now on we use the same notations as in the proof of Proposition 14.4.

Let $\varphi_0, \varphi_1, \ldots, \varphi_N$ be a basis of $H^0(V, \underline{O}(mK_V))$ and let $\underline{U}_k$, $k = 0, 1, \ldots, N$ be an affine covering of $\mathbb{P}^N$ such that $\underline{U}_i$ is an n-dimensional affine space with global coordinates $(u_k^1, u_k^2, \dots, u_k^N)$. We shall consider the complex manifold $\mathbb{P}^N \times \mathbb{K}$, where $\mathbb{K}$ is the total space of the canonical line bundle of V. $\mathbb{K}$ is covered by open subsets $\{V_i \times \mathbb{C}\}_{i\in I}$ with local coordinates $(z_i^1,\dots,z_i^n,w_i)$. On V_i, φ_0, $\varphi_1,\dots,\varphi_N$ can be represented in the form

$$\varphi_j = \varphi_{j,i}(z_i^1,\dots,z_i^n)(dz_i^1 \wedge \dots \wedge dz_i^n)^m,$$

where $\varphi_{j,i}$ is holomorphic on V_i. Let $\underline{V}$ be a subvariety of $\mathbb{P}^N \times \mathbb{K}$ defined by the equations

$$(w_i)^m = u_k^1\varphi_0 + u_k^2\varphi_2 + \dots + u_k^k\varphi_{k-1} + \varphi_k + u_k^{k+1}\varphi_{k+1} + \dots + u_k^N\varphi_N$$

in $\underline{U}_k \times V_i \times \mathbb{C}$. The natural projection of $\mathbb{P}^N \times \mathbb{K}$ onto $\mathbb{P}^N \times V$ induces a holomorphic surjective mapping $f : \underline{V} \longrightarrow \mathbb{P}^N \times V$. We set $g = p_1 \circ f : \underline{V} \longrightarrow \mathbb{P}^N$ where p_1 is the projection of $\mathbb{P}^N \times V$ onto $\mathbb{P}^N$. For any point $p \in \mathbb{P}^N$, there exists an element $\varphi \in H^0(V, \underline{O}(mK_V))$, unique up to a constant factor such that the variety $V_p = g^{-1}(p)$ is an m-fold branched covering of V constructed by $c\varphi$ in the proof of Proposition 14.4 for a suitable non-zero constant c. Conversely, if V' is an m-fold branched covering of V constructed by an element $\varphi \in H^0(V, \underline{O}(mK_V))$, then there exists a point $p \in \mathbb{P}^N$ such that V' is isomorphic to $V_p = g^{-1}(p)$. By our construction, $\underline{V}$ is algebraic, the morphism g is proper and every fibre of g is equi-dimensional.

Next we consider the situation that algebraic varieties X, S and a proper surjective equidimensional morphism $f : X \longrightarrow S$ are given. Let η be a generic point (in the sense of Grothendieck) of S and X_η be the fibre over η. X_η is an algebraic variety defined over a field $\mathbb{C}(S)$. By virtue of Hironaka [1], Main Theorem I, p.132 (see also Theorem 2.22), there exists a resolution of singularities $\pi_\eta : \tilde{X}_\eta \longrightarrow X_\eta$ of X_η defined over $\mathbb{C}(S)$. Therefore, there exist a complete algebraic variety $\tilde{X}$, a surjective morphism $\tilde{f} : \tilde{X} \longrightarrow S$ and a birational morphism $\pi : \tilde{X} \longrightarrow X$ such that for a generic point η of S, the fibre $\tilde{f}^{-1}(\eta)$ of $\tilde{f}$ over η is isomorphic to $\tilde{X}_\eta$ over $\mathbb{C}(S)$. As a consequence, there exists a Zariski open subset T_0 of S such that for any point $x \in T_0$, the fibre $\tilde{X}_x = \tilde{f}^{-1}(x)$ is a non-singular model of $X_x = f^{-1}(x)$. Let S_i, $i = 1,2,\dots,\ell$ be irreducible components of $S - T_0$ and $f_i : X_i \longrightarrow S_i$ be the pull back of the family $f : X \longrightarrow S$ over S_i. Then applying the same argument as above, we conclude that there exists a Zariski open subset T_i of S_i and a proper surjective morphism $\tilde{f}_i : \tilde{X}_i \longrightarrow T_i$ such that for any point $x \in T_i$, $\tilde{X}_{i,x} = \tilde{f}_i^{-1}(x)$ is a non-singular model of $X_{i,x} = f_i^{-1}(x)$. In this way, we can prove that there exist a finite number of (non-complete) algebraic varieties T_i, $\tilde{X}_i$, $i=0,1,\dots,\ell$, and proper surjective morphisms $\tilde{f}_i : \tilde{X}_i \longrightarrow T_i$ such that for any point $y \in S$, there exist T_i and a point $y_i \in T_i$ such that $\tilde{X}_{i,y_i} = \tilde{f}_i^{-1}(y_i)$ is a non-singular model of $X_y = f^{-1}(y)$. Furthermore, by the Thom-Mather isotopy theorem $\dim_{\mathbb{C}} H^k(\tilde{X}_{i,x}, \mathbb{C})$ is uniformly bounded for $x \in S$ (see Mather [1]).

We apply the above consideration to our family $g : \underline{V} \longrightarrow \mathbb{P}^N$. $\underline{V}$ may not be connected. However, applying the above argument to each connected component of $\underline{V}$, we conclude that for any $p \in \mathbb{P}^N$, there exists a non-singular model $\tilde{V}_p$ of $V_p = g^{-1}(p)$ such that $b_n(\tilde{V}_p)$ is uniformly bounded for any $p \in \mathbb{P}^N$. Q.E.D.

The above theorem was first proved by Nakamura and Ueno [1] under the assumption that $\Phi_{mK}(V)$ is not a ruled variety. The present proof is due to Deligne.

§15. Addition formula

In this section we shall prove the following theorem.

Theorem 15.1. Let $\pi : V \longrightarrow W$ be an analytic fibre bundle over a complex manifold W whose fibre and structure group are a Moishezon manifold F and the automorphism group $\mathrm{Aut}(F)$ of F, respectively. Then we have

$$\kappa(V) = \kappa(F) + \kappa(W).$$

Proof. Let $\{W_i\}_{i\in I}$ be a sufficiently fine finite open covering of W such that $(w_i^1,\dots,w_i^\ell)$ is a system of local coordinates of W_i. By $\underline{\mathrm{Aut}}(F)$ we denote the sheaf of germs of holomorphic sections of $\mathrm{Aut}(F)$ over W. The complex fibre bundle $\pi : V \longrightarrow W$ is determined by a 1-cocycle $\{F_{ij}\}$, where $F_{ij} \in H^0(W_i \cap W_j, \underline{\mathrm{Aut}}(F))$. Let $\{U_j\}_{j\in J}$ be a sufficiently fine finite open covering of F such that $(u_j^1,\dots, u_j^n)$ is a system of local coordinates of U_j. The fibre bundle V is covered by an open covering $\{V_{ij}\}$ where V_{ij} is analytically isomorphic to $W_i \times U_j$. The transition functions $\{K_{(i,j)(k,\ell)}(V)\}$ of the canonical line bundle K_V of V are given by

$$K_{(i,j)(k,\ell)}(V)=\left(\det\left(\frac{\partial(w_i^1,\dots,w_i^\ell,F_{ik}^1(u_k,z_\ell),\dots,F_{ik}^n(u_k,z_\ell))}{\partial(w_k^1,\dots,w_k^\ell,z_\ell^1,\dots\dots\dots,z_\ell^n)}\right)\right)^{-1}$$

$$= \det\left(\frac{\partial(w_i^1,\dots,w_i^\ell)}{\partial(w_k^1,\dots,w_k^\ell)}\right)^{-1}\cdot \det\left(\frac{\partial(F_{ik}^1(u_k,z_\ell),\dots,F_{ik}^n(u_k,z_\ell))}{\partial(z_\ell^1,\dots\dots\dots,z_\ell^n)}\right)^{-1}.$$

Hence we have

$$K_V = \pi^*(K_W)\otimes L,$$

where L is a line bundle determined by transition functions

$$(15.2) \qquad \{ \det \left(\frac{\partial(F^1_{ik}(u_k, z_\ell), \ldots, F^n_{ik}(u_k, z_\ell))}{\partial(z^1_\ell, \ldots\ldots\ldots, z^n_\ell)} \right)^{-1} \} \quad .$$

If we restrict the line bundle L to the fibre $V_w = \pi^{-1}(w)$, $w \in W$, then $L_{|V_w}$ is nothing other than the canonical line bundle K_{V_w}.

By $\pi_*(K_V^{\otimes m})$ and $\pi_*(L^{\otimes m})$ we denote the vector bundle associated with locally free sheaves $\pi_*\underline{O}(K_V^{\otimes m})$ and $\pi_*\underline{O}(L^{\otimes m})$. With this notation we have

$$\pi_*(K_V^{\otimes m}) = K_W^{\otimes m} \otimes \pi_*(L^{\otimes m}) .$$

On W_i, $\pi_*(L^{\otimes m})_{|W_i}$ is analytically isomorphic to a trivial vector bundle $W_i \times H^0(F, \underline{O}(mK_F))$. From 15.2 we can easily show that transition functions $\{G_{ij}\}$ of this vector bundle $\pi_*(L^{\otimes m})$ are given by

$$G_{ij} = \rho_m(F_{ij}),$$

where ρ_m is the pluricanonical representation of $\mathrm{Aut}(F)$ into $GL(H^0(F, \underline{O}(mK_F)))$.

Let A_m be the subgroup of $\rho_m(\mathrm{Aut}(F))$ generated by $\rho_m(F_{ij})$ for all $(i, j) \in I \times J$. By Theorem 14.10, A_m is a finite group. Hence there exists a finite unramified covering manifold $f : \tilde{W} \longrightarrow W$ such that the induced vector bundle $f^*(\pi_*(K_V^{\otimes m}))$ is isomorphic to $K_{\tilde{W}}^{\otimes m} \otimes H^0(F, \underline{O}(mK_F))$. Therefore, we infer that $\kappa(V) = -\infty$ follows from $\kappa(W) = -\infty$ or $\kappa(F) = -\infty$.

If $\kappa(W) \neq -\infty$ and $\kappa(F) \neq -\infty$, we let $\tilde{f} : (\tilde{V}, \tilde{\pi}, \tilde{W}) \longrightarrow (V, \pi, W)$ be the lifting of $\pi : V \longrightarrow W$ over $\tilde{W}$. Note that

$$\begin{aligned} H^0(\tilde{V}, \underline{O}(mK_{\tilde{V}})) &= H^0(\tilde{W}, \tilde{\pi}_*\underline{O}(mK_{\tilde{V}})) \\ &= H^0(\tilde{W}, \tilde{\pi}_* f^*\underline{O}(mK_V)) \\ &= H^0(\tilde{W}, f^* \pi_*\underline{O}(mK_V)). \end{aligned}$$

Then, combining this with

$$H^0(W,\ f^*\pi_*\underline{O}(mK_V)) = H^0(W,\ \underline{O}(mK_W)) \otimes H^0(F,\ \underline{O}(mK_F)) \ ,$$

we obtain

$$H^0(\tilde{V},\ \underline{O}(mK_{\tilde{V}})) = H^0(\tilde{W},\ \underline{O}(mK_{\tilde{W}})) \otimes H^0(F,\ \underline{O}(mK_F)).$$

From this, it follows that $\kappa(\tilde{V}) = \kappa(\tilde{W}) + \kappa(F)$. On the other hand by Theorem 6.10, 2) we have $\kappa(V) = \kappa(\tilde{V})$ and $\kappa(W) = \kappa(\tilde{W})$. Hence we have

$$\kappa(V) = \kappa(W) + \kappa(F). \qquad \text{Q.E.D.}$$

<u>Remark 15.3</u>. Let T and g be the three-dimensional complex torus and the automorphism of the torus constructed in Remark 14.6. Let E be an elliptic curve with fundamental periods $\{1, \omega\}$. Let G be a free abelian group of analytic automorphisms of $\mathbb{C} \times T$ generated by two automorphisms :

$$g_1 : (z,\ q) \longmapsto (z+1,\ q)$$

$$g_2 : (z,\ q) \longmapsto (z+\omega,\ g(q)) \ .$$

The group G acts on $\mathbb{C} \times T$ properly discontinously and freely. The quotient manifold $V = \mathbb{C} \times T/G$ is a fibre bundle over E whose fibre and structure group are T and Aut(T), respectively. Remark 14.6 and Example 6.9,2) imply that $\kappa(V) = -\infty$.

Therefore, <u>the addition formula does not hold in general without the assumption that a fibre F is a Moishezon manifold.</u>

Chapter VII

Examples of complex manifolds

In this chapter, we shall provide examples of complex manifolds which show that several new phenomena occur if dimensions of complex manifolds are strictly bigger than two.

In §16, we shall mainly study Kummer manifolds (see Definition 16.1). Kodaira dimensions of Kummer manifolds are always non-positive. We shall construct several examples of Kummer manifolds of parabolic type and show that

1) there exists a three-dimensional Kummer manifold V with a trivial canonical bundle such that $\dim_{\mathbb{C}} H^1(V, \Theta) = 0$ (see Example 16.12 and Lemma 16.12.4 ; note that $\dim H^1(S, \Theta) > 0$ for any surface S with a trivial canonical bundle) :

2) there exists a three-dimensional Kummer manifold V of parabolic type such that, for any bimeromorphically equivalent model V^* of V, the pluricanonical bundle $mK(V^*)$ is not analytically trivial (see Corollary 16.11.3, Corollary 16.12.3 and Proposition 16.17).

We shall also show that

3) there exists a three-dimensional algebraic manifold V of hyperbolic type such that, for any bimeromorphically equivalent model V^* of V, the pluricanonical system $|mK(V^*)|$ has always fixed components for any positive integer m (see Example 16.18).

In this section, we shall also show that Conjecture K_n is true for Kummer manifolds of parabolic type.

The present section is taken from Ueno [3], I, §7 and Ueno [7].

In §17, we shall study the structure of three-dimensional parallelizable manifolds (see Definition 17.1) and their deformations. We shall show that

4) there exist a one parameter family $\{M_t\}_{|t|<\varepsilon}$ of three-dimensional complex manifolds such that $M = M_0$ is parallelizable and the invariants P_m, κ, q, r, t, g_k, $h^{p,q}$ are not invariant under this deformation (see Example 17.8).

Note that all complex manifolds M_t are non-Kähler. It is not yet known whether these invariants are invariant under (global) Kähler deformations or not. Moreover, we shall show that the deformation of the above parallelizable manifold M is obstracted and the deformation space of the manifold M has many irreducible components.

This section is taken from Nakamura [1].

In §18, we shall introduce many complex structures on products of of two homotopy spheres. Here we can find the deep connection between differential topology and theory of complex manifolds. Recently, Morita [1] has classified topologically all complex structures on $S^1 \times \Sigma^{2n-1}$ where Σ^{2n-1} is a (2n-1)-dimensional homotopy sphere which bounds a parallelizable manifold (see Theorem 18.4.2, below). It would be very interesting to study the relationship between deformations of such complex structures and Morita's topological classification. Moreover, in this section, we shall show that

5) there exists a complex manifold M which is diffeomorphic to $S^1 \times S^{2n-1}$, $n \geqq 3$, such that, for the algebraic reduction $\Psi^* : M^* \longrightarrow V$ of the manifold M, we cannot take the manifold M itself as M^* (see Corollary 18.1.3 and Remark 18.1.4).

The results in this section are due to several mathematicians. The references will be found in the text.

§16. Kummer manifolds

Let A be an abelian variety and let G be the cyclic group of order two of analytic automorphisms of A generated by the automorphism

$$\begin{array}{ccc} g : A & \longrightarrow & A \\ \cup & & \cup \\ z & \longmapsto & -z \end{array} \quad .$$

Classically, the quotient space A/G is called a Kummer variety. We generalize this notion as follows.

Definition 16.1. An algebraic manifold V is called a Kummer manifold if there exist an abelian variety A and a finite group G of analytic automorphisms of A such that V is a non-singular model of the quotient variety A/G.

An algebraic manifold V is called a generalized Kummer manifold if there exist an abelian variety A and a generically surjective rational mapping $f : A \longrightarrow V$ of A onto V (note that we do not assume that dim A = dim V).

By global coordinates of an abelian variety A, we mean global coordinates of the universal covering of $\mathbb{C}^n$ of A such that, in terms of these coordinates, covering transformations are represented by translations by element of a lattice Δ in $\mathbb{C}^n$, where $A = \mathbb{C}^n/\Delta$.

Theorem 16.2. For a generalized Kummer manifold V, we have

$$\kappa(V) \leqq 0 .$$

Proof. Let $f : A \longrightarrow V$ be a generically surjective rational mapping of an ablian variety A onto V. We can assume that, at the origin 0 of A, f is holomorphic and of maximal rank.

Then there are global coordinates $z_1, z_2, \ldots, z_n$ of A such that $z_1, z_2, \ldots, z_\ell$ induce local coordinates of a small neighbourhood W of $f(0)$ in V with center $f(0)$.

Let φ be an element of $H^0(V, \underline{O}(mK_V))$. We can represent φ on W in the form

(16.2.1). $\varphi = \Psi(z_1, z_2, \ldots, z_\ell)(dz_1 \wedge \cdots \wedge dz_\ell)^m$,

where Ψ is holomorphic on W. Let U be a small neighbourhood of the origin 0 in A such that $f(U) \subset W$. The pull back $f^*(\varphi)$ can be written in the same form 16.2.1 on U. Taking the exterior product on both sides of 16.2.1 by $(dz_{\ell+1} \wedge \cdots \wedge dz_n)^m$, we have

(16.2.2) $(f^*\varphi) \wedge (dz_{\ell+1} \wedge \cdots \wedge dz_n)^m = \Psi(z_1, z_2, \ldots, z_\ell)(dz_1 \wedge \cdots \wedge dz_\ell \wedge$

$\wedge dz_{\ell+1} \wedge \cdots \wedge dz_n)^m$,

on U. On the other hand, by an argument similar to that of the proof of Lemma 6.3, $(f^*\varphi) \wedge (dz_{\ell+1} \wedge \cdots \wedge dz_n)^m$ is an element of $H^0(A, \underline{O}(mK_A))$. Therefore, by 16.2.2, $f^*\varphi \wedge (dz_{\ell+1} \wedge \cdots \wedge dz_n)^m$ is written in the form

$$c(dz_1 \wedge \cdots \wedge dz_n)^m,$$

where c is a constant. Hence the holomorphic function Ψ must be constant. This implies that any two elements of $H^0(V, \underline{O}(mK_V))$ are linearly dependent. Q.E.D.

Remark 16.3. Theorem 16.2 is also true for a compact complex manifold M which is an image of a generically surjective meromorphic mapping $f : T \longrightarrow M$ of a complex torus T onto M. The proof is the same as above.

Corollary 16.4. For a generalized Kummer manifold V, the Albanese mapping $\alpha : V \longrightarrow A(V)$ is surjective. Hence, a fortiori, we have

$$t(V) = q(V) \leqq \dim V .$$

Proof. A non-singular model of the image $\alpha(V)$ of the Albanese mapping α is also a generalized Kummer manifold. Therefore, we have $\kappa(\alpha(V)) \leqq 0$. Hence the corollary follows from Corollary 10.6.

Q.E.D.

Proposition 16.5. Let V be a generalized Kummer manifold.

1) If $q(V) = \dim V$, the Albanese mapping $\alpha : V \longrightarrow A(V)$ is a birational morphism.

2) Any fibre of the Albanese mapping $\alpha : V \longrightarrow A(V)$ is connected. Furthermore, general fibres of α are generalized Kummer manifolds.

The proof can be found in Ueno [3], I, §7.

Proposition 16.6. Let V be a generalized Kummer manifold of dimension ℓ. If $\kappa(V) = 0$, there exist an ℓ-dimensional abelian variety A and a generically surjective rational mapping $f : A \longrightarrow V$ of A onto V.

This was first proved by Ueno [3], I, §7. The following simple proof is due to K. Akao.

By definition there exist an abelian variety A and a generically surjective rational mapping $f : A \longrightarrow V$ of A onto V. By Theorem 2.23 there exists a modification $g : \tilde{A} \longrightarrow A$ of an algebraic manifold $\tilde{A}$ onto A such that $\tilde{A}$ is obtained from A by a succession of monoidal transformations with non-singular centers and $\tilde{f} = f \circ g : \tilde{A} \longrightarrow V$ is a morphism. Let $\mathcal{E}$ be the exceptional divisors which appear as a result of these monoidal transformations. There exists an open dense set V' in V such that $\tilde{f}$ is of maximal rank at any point of $\tilde{f}^{-1}(V')$. We choose a general point $x \in V'$. $\tilde{A}_x = \tilde{f}^{-1}(x)$ is disjoint union of algebraic manifolds. Let $\tilde{L}$ be one of irreducible components of $\tilde{A}_x$ which is not contained in $\mathcal{E}$. We set $L = g(\tilde{L})$. $\tilde{L}$ and L are birationally equivalent. We choose a point $p \in L$ such that, at p, f is holomorphic and of maximal rank. Then there exist global coordinates $z_1, z_2, \ldots, z_n$ of A such that $z_1 - a_1, z_2 - a_2, \ldots, z_\ell - a_\ell$ induce local coordinates in a neighbourhood W of $f(p)$ in V with center $f(p)$, where $p = [a_1, a_2, \ldots, a_n]$. As $\kappa(V) = 0$, there exists a non-zero section $\varphi \in H^0(V, \underline{O}(mK_V))$ for a suitable m. By the proof of Theorem 16.2, φ is expressed on W in the form

$$\varphi = c(dz_1 \wedge \cdots \wedge dz_\ell)^m,$$

where c is a constant. We choose a neighbourhood U of p in A such that $f(U) \subset W$. The pull back $f^*\varphi$ can be expressed on U in the form

$$f^*\varphi = c(dz_1 \wedge \cdots \wedge dz_\ell)^m .$$

Suppose that there exist two linearly independent holomorphic $(n-\ell)$-forms ω_1, ω_2 on A such that $g^*\omega_1$ and $g^*\omega_2$ induce linearly independent holomorphic $(n-\ell)$-forms on $\tilde{L}$. Since $g_{|\tilde{L}} : \tilde{L} \longrightarrow A$ induces a biholomorphic mapping between a neighbourhood of $g^{-1}(p)$ in $\tilde{L}$ and a neighbourhood $U \cap L$ of p in L and since local coordinates in $U \cap L$ are given by $z_{\ell+1} - a_{\ell+1}, \ldots, z_n - a_n$, $(\omega_1)^m \wedge f^*\varphi$ and $(\omega_2)^m \wedge f^*\varphi$ are linearly independent. But $(\omega_1)^m \wedge f^*\varphi$ and $(\omega_2)^m \wedge f^*\varphi$ are linearly dependent elements of $H^0(A, \underline{O}(mK_A))$. This is a contradicicition. Hence there exists only one holomorphic $(n - \ell)$-form ω on A such that $g^*\omega$ induces a holomorphic $(n - \ell)$-form on $\tilde{L}$. From Corollary 10.4, it follows that L is non-singular and a translation of an abelian subvariety A_1 of A by an element of A. If necessary, by taking a finite unramified covering of A, we can assume that $A = A_1 \times A_2$, where A_2 is another abelian subvariety of A.

The above argument shows that we can choose $z_1, \ldots, z_\ell$ as global coordinates of A_2 and $z_{\ell+1}, \ldots, z_\ell$ as global coordinates of A_1. The above point p can be written as (p_1, p_2), where $p_1 \in A_1$, $p_2 \in A_2$. Again, by the above argument, the composition

$$\begin{array}{ccccc} f_2 : A_2 & \hookrightarrow & A & \xrightarrow{f} & V \\ \cup\!\!\!\!\!\!\;\omega & & \cup\!\!\!\!\!\!\;\omega & & \\ z & \longmapsto & (p_1, z) & & \end{array}$$

is a well defined meromorphic mapping.

Moreover, in a neighbourhood of $p_2 \in A_2$, f_2 is surjective. Therefore, the meromorphic mapping $f_2 : A_2 \longrightarrow V$ is surjective. As $\dim A_2 = \dim V = \ell$, we obtain the desired result. Q.E.D.

Now we shall prove the following theorem which has been already mentioned in (11.4).

<u>Theorem 16.7</u>. Let V be a Kummer manifold. Then $\kappa(V) = 0$ if and only if there exists an algebraic manifold V^* such that

1) V^* is birationally equivalent to V ;
2) the Albanese mapping $\alpha : V^* \longrightarrow A(V^*)$ has the structure of an analytic fibre bundle whose fibre is a Kummer manifold of parabolic type.

<u>Proof</u>. The "if" part is a consequence of Theorem 15.1. We shall prove the "only if" part. There exist an Abelian variety A and a finite group G of analytic automorphisms of A such that V is a non-singular model of the quotient variety A/G and $f : A \longrightarrow A/G$ is the quotient morphism. Let $g : \tilde{A} \longrightarrow A$ be a finite unramified covering of A such that $\tilde{A}$ is a product of simple abelian varieties. We let S be a singular locus of the quotient space A/G. The algebraic set $f^{-1}(S)$ is of codimension at least two in A. Hence

$\tilde{S} = g^{-1}(f^{-1}(S))$ is of codimension at least two in $\tilde{A}$. We set $\tilde{A}' = \tilde{A} - \tilde{S}$ and $V' = A/G - S$. From the proof of Theorem 7.2, since $\kappa(V) = 0$, we infer that $f \circ g|_{\tilde{A}'} : \tilde{A}' \longrightarrow V'$ is a finite unramified covering. Hence the fundamental group $\pi_1(A')$ is a subgroup of the fundamental group $\pi_1(V')$ of <u>finite index</u>. Therefore, there exists a subgroup H of $\pi_1(A')$ such that H is a <u>normal subgroup</u> of $\pi_1(V')$ of <u>finite index</u>. As the algebraic set $\tilde{S}$ is at least of codimension two, we have $\pi_1(\tilde{A}) = \pi_1(\tilde{A}')$. Hence there exists a finite unramified covering $u : \tilde{\tilde{A}} \longrightarrow \tilde{A}$ which corresponds to the subgroup $H \subset \pi_1(\tilde{A})$. As $\tilde{A}$ is a product of simple abelian varieties, $\tilde{\tilde{A}}$ is a product of simple abelian varieties. We set $\tilde{\tilde{A}}' = u^{-1}(\tilde{A}')$. Then $f \circ g \circ u|_{\tilde{\tilde{A}}'} : \tilde{\tilde{A}}' \longrightarrow V'$ is a finite Galois covering. Hence the field extension $\mathbb{C}(\tilde{\tilde{A}})/\mathbb{C}(V)$ induced by this morphism is Galois. The Galois group $\tilde{G}$ of this field extension is a group of birational transformations of $\tilde{\tilde{A}}$. In view of Lemma 9.11, $\tilde{G}$ is a group of analytic automorphisms of $\tilde{\tilde{A}}$. As we have $\mathbb{C}(\tilde{\tilde{A}})^{\tilde{G}} = \mathbb{C}(V)$, V is a non-singular model of a quotient variety $\tilde{\tilde{A}}/\tilde{G}$. Therefore, we can assume that A itself is already a product of simple abelian varieties.

Let $\bar{f} : A \longrightarrow V$ be a generically surjective rational mapping obtained from the quotient morphism $A \longrightarrow A/G$. From Proposition 9.24, we have

$$q(V) = g_1(V) = \dim H^0(A, \Omega_A^1)^G .$$

Hence there exist global coordinates $z_1, z_2, \ldots, z_\ell$ of A such that $dz_{\ell-q+1}, dz_{\ell-q+2}, \ldots, dz_\ell$ are invariant under the action of the group G. We can write $A = A_1 \times A_2$ where A_1 and A_2 are abelian subvariety of A such that A_2 is a finite unramified covering of $A(V)$ by the morphism $\alpha \circ \bar{f}$. (We can assume that $\alpha \circ \bar{f}$ is a group homomorphism.) In this situation, we may consider that A_1 has global coordinates $z_1, z_2, \ldots, z_{\ell-q}$ and A_2 has global coordinates $z_{\ell-q+1}$, $z_{\ell-q+2}, \ldots, z_\ell$.

In terms of these coordinates, an analytic automorphism $g \in G$ is expressed in the form

$$g \quad : \quad \begin{pmatrix} z_1 \\ z_2 \\ \vdots \\ z_\ell \end{pmatrix} \longmapsto A(g) \cdot \begin{pmatrix} z_1 \\ z_2 \\ \vdots \\ z_\ell \end{pmatrix} + \begin{pmatrix} \underline{a}_1(g) \\ \underline{a}_2(g) \end{pmatrix},$$

where

$$A(g) = \begin{pmatrix} \overbrace{A_1(g)}^{\ell-q} & \overbrace{B(g)}^{q} \\ 0 & I_q \end{pmatrix} \begin{matrix} \}\ell-q \\ \}\ q \end{matrix} \in \quad GL(\ell, \mathbb{C}) ,$$

$$\underline{a}_1(g) = \begin{pmatrix} a_1(g) \\ a_2(g) \\ \vdots \\ a_{\ell-q}(g) \end{pmatrix} \in \mathbb{C}^{\ell-q}, \quad \underline{a}_2(g) = \begin{pmatrix} a_{\ell-q+1}(g) \\ a_{\ell-q+2}(g) \\ \vdots \\ a_\ell(g) \end{pmatrix} \in \mathbb{C}^q .$$

We set $H = \{ g \in G \mid \underline{a}_2(g) = 0 \}$. Then H is a normal subgroup of G.

We set

$$C \ = \ - \frac{1}{|H|} \sum_{h \in H} B(h).$$

Since we have

$$B(g \circ h) \ = \ A_1(g) \cdot B(h) \ + \ B(g) \ ,$$

it follows that

(16.7.1) $\quad A_1(g) \cdot C - C \ = \ B(g) \ , \quad g \in H$.

Let $w_1, w_2, \ldots, w_{\ell-q}, z_{\ell-q+1}, \ldots, z_1$ be new global coordinates of A defined by

$$\begin{pmatrix} w_1 \\ w_2 \\ w_{\ell-q} \end{pmatrix} = (I_{\ell-q}, \; C) \cdot \begin{pmatrix} z_1 \\ z_2 \\ \vdots \\ z_\ell \end{pmatrix}$$

In view of 16.7.1, in terms of these new coordinates, any automorphism $h \in H$ is written in the form

$$(16.7.2) \qquad \begin{pmatrix} w \\ z \end{pmatrix} \longmapsto \begin{pmatrix} A_1(h) & 0 \\ 0 & I_q \end{pmatrix} \begin{pmatrix} w \\ z \end{pmatrix} + \begin{pmatrix} \underline{a}_1(h) \\ 0 \end{pmatrix},$$

where

$$w = \begin{pmatrix} w_1 \\ w_2 \\ \vdots \\ w_{\ell-q} \end{pmatrix}, \quad z = \begin{pmatrix} z_{\ell-q+1} \\ z_{\ell-q+2} \\ \vdots \\ z_\ell \end{pmatrix}.$$

Note that for any <u>fixed</u> z, $w_1, w_2, \ldots, w_{\ell-q}$ are global coordinates of A_1.

Let H_1 be the group of analytic automorphisms of the abelian variety A_1 consisting of all automorphisms

$$h_1 : \begin{pmatrix} z_1 \\ z_2 \\ \vdots \\ z_{\ell-q} \end{pmatrix} \longmapsto A_1(h) \cdot \begin{pmatrix} z_1 \\ z_2 \\ \vdots \\ z_{\ell-q} \end{pmatrix} + \underline{a}_1(h), \quad h \in H.$$

We set $F = A_1/H_1$. The natural morphism $\tau : X = A/H \longrightarrow A_2$, induced by the projection $A = A_1 \times A_2 \longrightarrow A_2$ has the structure of a fibre bundle whose fibre is the quotient space $F = A_1/H_1$, in view of 16.7.2. As H is a normal subgroup of G, the quotient group $\bar{G} = G/H$ operates on X. Any element $\bar{g} \neq id$ of $\bar{G}$ is a fibre preserving analytic automorphism of the fibre space $\tau : X \longrightarrow A_2$ and has no fixed points, since $\underline{a}_2(g) \neq 0$. Moreover, $\bar{g}$ operates on A_2 by

$$\begin{pmatrix} z_{\ell-q+1} \\ z_{\ell-q+2} \\ \vdots \\ z_{\ell} \end{pmatrix} \longmapsto \begin{pmatrix} z_{\ell-q+1} \\ z_{\ell-q+2} \\ \vdots \\ z_{\ell} \end{pmatrix} + \underline{a}_2(g) ,$$

where $g \in G$ is a representative of $\bar{g}$. Therefore, the morphism $\tau : X \longrightarrow A_2$ is $\bar{G}$ equivariant.

By Theorem 2.22, there exists a non-singular model F^* of F such that $\mathrm{Aut}(F)$ can be lifted to a group of analytic automorphisms of F^*. On replacing the fibre F of the fibre bundle $\tau : X \longrightarrow A_2$ by F^*, we obtain the associated fibre bundle $\tau^* : X^* \longrightarrow A_2$. The group $\bar{G}$ operates on X^* and the morphism τ^* is $\bar{G}$ equivariant. Furthermore, any element $\bar{g} \neq \mathrm{id}$ operates on X^* without fixed points. Thus, the quotient $\pi : V^* = X^*/\bar{G} \longrightarrow A_2/\bar{G}$ is a fibre bundle over $A_2/\bar{G}$ whose fibre is F^*. The algebraic manifold V^* is birationally equivalent to V. Because of our construction, it is easy to show that

$$A_2/\bar{G} = A(V^*) = A(V),$$

and that the morphism $\pi : V^* \longrightarrow A_2/\bar{G}$ is the Albanese mapping.

Q.E.D.

<u>Remark 16.8</u>. The above fibre bundle $\alpha : V^* \longrightarrow A(V^*)$ is not only an analytic fibre bundle but also <u>a fibre bundle in the etale topology</u>. The proof is left to the reader (see Ueno [3], I, Remark 7.16).

The proof of the following Theorem can be found in Ueno [3], I, Theorem 7.17. The proof is based on the argument given in 11.9.

<u>Theorem 16.9</u>. Let V be a generalized Kummer manifold. Suppose that $q(V) = \dim V - 1$. Then, in this situation, we have the following :

1) $\kappa(V) = -\infty$ if and only if general fibres of the Albanese mapping $\alpha : V \longrightarrow A(V)$ are $\mathbb{P}^1$;

2) $\kappa(V) = 0$ if and only if the Albanese mapping $\alpha : V \longrightarrow A(V)$ is birationally equivalent to an analytic fibre bundle over $A(V)$ whose fibre is an elliptic curve.

Now we shall provide examples of Kummer manifolds. For that purpose, first, we shall resolve certain singularities.

(16.10) <u>The canonical resolution of the singular point of</u> Q^n_m .

Let G be the cyclic group of analytic automorphisms of $\mathbb{C}^n$ generated by an automorphism

$$g : (z_1, z_2, \ldots, z_n) \longmapsto (e_m z_1, e_m z_2, \cdots, e_m z_n) ,$$

where $e_m = \exp(2\pi\sqrt{-1}/m)$. The quotient space $Q^n_m = \mathbb{C}^n/G$ has only one sigular point p, which corresponds to the origin of $\mathbb{C}^n$.

Let U_i, $i = 1, 2, \ldots, n$ be n copies of $\mathbb{C}^n$ whose global coordinates are $(w^1_i, w^2_i, \ldots, w^n_i)$, respectively. We shall construct a complex manifold $M = \bigcup_{i=1}^{n} U_i$ by identifying open subsets of U_{i-1} and U_i, $i = 1, 2, \ldots, n$, through the following relations :

$$\begin{cases} w^k_i = \dfrac{w^k_{i-1}}{w^i_{i-1}}, \quad k \neq i-1, i, \\[2ex] w^{i-1}_i = \dfrac{1}{w^i_{i-1}} \\[2ex] w^i_i = (w^i_{i-1})^m \cdot w^{i-1}_{i-1} \end{cases} .$$

Meromorphic mappings

$$T_{U_i} : \mathbb{C}^n \longrightarrow U_i, \quad i = 1, 2, \ldots, n$$

$$\omega \qquad\qquad \omega$$

$$(z_1, z_2, \ldots, z_n) \longmapsto \left(\frac{z_1}{z_i}, \ldots, \frac{z_{i-1}}{z_i}, (z_i)^m, \frac{z_{i+1}}{z_i}, \ldots, \frac{z_n}{z_i}\right)$$

induce a meromorphic mapping $T : Q^n_m \longrightarrow M$. Let E be the submanifold of M defined by the equations

$$w^i_i = 0, \quad \text{in } U_i, \quad i = 1, 2, \ldots, n.$$

E is analytically isomorphic to an $(n-1)$-dimensional complex projective space $\mathbb{P}^{n-1}$. The meromorphic mapping $T : Q^n_m \longrightarrow M$ induces an isomorphism between $Q^n_m - p$ and $M - E$. It follows that M is a non-singular model of Q^n_m.

We set $V = \{(z_1, \ldots, z_n) \in \mathbb{C}^n \mid |z_i| < (\varepsilon)^{\frac{1}{m}}\}$.

The group G operates on V and the quotient space V/G is a neighbourhood of p in Q^n_m. Let $\tilde{M}$ be an open set in M defined by inequalities :

$$|(w^k_i)^m \cdot w^i_i| < \varepsilon \;, \quad k \neq i \;,$$

$$|w^i_i| < \varepsilon \;, \quad \text{in } U_i, \quad i = 1, 2, \ldots, n \;.$$

Then $E \subset \tilde{M}$ and the meromorphic mapping T induces an isomorphism between $V/G - p$ and $\tilde{M} - E$. Hence $\tilde{M}$ is a non-singular model of the quotient space V/G.

This procedure of resolving singularities is called the <u>canonical resolution</u> of the singularity of Q^n_m.

<u>Example 16.11</u>. Let T be a complex torus of dimension n and let G be the cyclic group of analytic automorphisms of T generated by an automorphism

$$\begin{array}{cccc} g : & T & \longrightarrow & T \\ & \omega & & \omega \\ & [z_1, z_2, \ldots, z_n] & \longmapsto & [-z_1, -z_2, \ldots, -z_n]. \end{array}$$

The automorphism g has 2^{2n} fixes points. Hence, if $n \geq 2$, the quotient space T/G has 2^{2n} singular points which correspond to the fixed points. Each singular point has a neighbourhood which is isomorphic to a neighbourhood of the singular point p in Q_2^n. From the canonical resolution of its singularities (16.10), we obtain a non-singular model $K^{(n)}$ of T/G. If $n = 1$, $K^{(1)} = T/G$ is $\mathbb{P}^1$.

Lemma 16.11.1. For $n \geq 2$, we have

$$g_k(K^{(n)}) = \begin{cases} \binom{n}{k}, & k \equiv 0 \ (2), \\ 0, & k \equiv 1 \ (2), \end{cases} \qquad k = 1, 2, \ldots, n,$$

$$P_m(K^{(n)}) = \begin{cases} 1, & mn \equiv 0 \ (2), \\ 0, & mn \equiv 1 \ (2), \end{cases} \qquad m = 1, 2, \ldots .$$

Hence $\kappa(K^{(n)}) = 0$, if $n \geqq 2$. Moreover for $mn \equiv 0$ (2), the m-th canonical divisor has the form

$$\sum_{i=1}^{2^{2n}} m\left(\frac{n}{2} - 1\right)E_i ,$$

where $E_i \xrightarrow{\sim} \mathbb{P}^{n-1}$ appears in the canonical resolution of the singularities of A/G.

Proof. As we have

$$\dim H^0(T, \Omega_T^k)^G = \begin{cases} \binom{n}{k}, & k \equiv 0 \ (2) \\ 0, & k \equiv 1 \ (2), \end{cases}$$

the first part of the lemma holds in view of Proposition 9.24. On the other hand, we have

$$\dim H^0(T, \underline{O}(mK_T))^G = \begin{cases} 1, & mn \equiv 0 \ (2) \\ 0, & mn \equiv 1 \ (2) \end{cases} .$$

We use the notations of 16.10 freely. For $mn \equiv 0$ (2), we have

$$(dz_1 \wedge \cdots \wedge dz_n)^m = \frac{(w_i^1)^{m(n-2)/2}}{2^m} \cdot (dw_i^1 \wedge \cdots \wedge dw_i^n)^m .$$

This implies that G-invariant m-tuple n-form $(dz_1 \wedge \cdots \wedge dz_n)^m$ induces an element of $H^0(K^{(n)}, \underline{O}(mK))$. Q.E.D.

Lemma 16.11.2. $K^{(n)}$ is simply connected.

This is proved by Spanier [1].

Corollary 16.11.3. If $n \equiv 1$ (2), there does not exist a complex manifold V such that V is bimeromorphically equivalent to $K^{(n)}$ and $2K_V$ is analytically trivial.

Since $p_g(K^{(n)}) = 0$ and $K^{(n)}$ is simply connected, this is a consequence of the following lemma.

Lemma 16.11.4. Let F be a line bundle over a complex manifold M. Suppose that there exists a positive integer m such that $F^{\otimes m}$ is analytically trivial but $F^{\otimes \ell}$ is not analytically trivial for $\ell = 1, 2, \ldots, m-1$. Then there exists an m-fold unramified covering manifold $\tilde{M}$ of M.

Proof. Let $\{U_i\}_{i \in I}$ be an open covering of M and let $\{f_{ij}\}$ be transition functions of F with respect to this covering. As $F^{\otimes m}$ is analytically trivial, there exist non-vanishing holomorphic functions g_i on U_i, $i = 1, 2, \ldots,$ such that

$$g_i = (f_{ij})^m g_j .$$

Let $\mathbb{F}$ be a total space of the line bundle F. $\mathbb{F}$ is covered by open sets $U_i \times \mathbb{C}$. Let $\tilde{M}$ be the submanifold of $\mathbb{F}$ defined by equations

$$(z_i)^m = g_i, \quad \text{in} \quad U_i \times \mathbb{C} ,$$

where z_i is a coordinate of $\mathbb{C}$. Then it is clear that $\tilde{M}$ is an m-fold unramified covering of M. Q.E.D.

Lemma 16.11.5. Let Θ be the sheaf of germs of holomorphic vector fields on $K^{(n)}$. We have

$$\dim H^1(K^{(n)}, \Theta) = \begin{cases} 20, & n = 2 \\ n^2, & n \geqq 3. \end{cases}$$

When $n = 2$, $K^{(2)}$ is a K3 surface and the result is well known (see for example Kodaira [3], I, p. 782). When $n \geqq 3$, this is proved by Ueno [3], I, Lemma 8.5. If $K^{(n)}$ is algebraic, the proof was first given by Schlessinger [1].

Example 16.12. Let $E = E_\rho$ be the elliptic curve with period matrix $(1, \rho)$, where $\rho = \exp(2\pi\sqrt{-1}/3)$.
We set $E_\rho^n = \underbrace{E \times E \times \cdots \times E}_{n}$. Let G be the cyclic group of order three of analytic automorphisms of E_ρ^n generated by the automorphism

$$\begin{array}{cccc} g & : & E_\rho^n & \longrightarrow & E_\rho^n \\ & & \cup\!\!\!\!\!\!\;\;\; \omega & & \cup\!\!\!\!\!\!\;\;\; \omega \\ & & (z_1, z_2, \ldots, z_n) & \longmapsto & (\rho z_1, \rho z_2, \ldots, \rho z_n) \end{array} .$$

The automorphism g has 3^n fixed points. Hence, if $n \geqq 2$, the quotient space E_ρ^n/G has 3^n singular points corresponding to the fixed points of g. Each singular points has a neighbourhood in E_ρ^n/G which is analytically isomorphic to a neighbourhood of the singular point p in Q_3^n. By the canonical resolution of singularities 16.10, we obtain a non-singular model $L^{(n)}$ of E_ρ^n/G.
If $n = 1$, $L^{(1)} = E_\rho/G$ is $\mathbb{P}^1$.

Lemma 16.12.1.

$$g_k(L^{(n)}) = \begin{cases} \binom{n}{k}, & k \equiv 0 \ (3), \\ 0, & k \not\equiv 0 \ (3), \end{cases} \qquad k = 1, 2, \ldots, n .$$

$$P_m(L^{(n)}) = \begin{cases} 1, & n \geqq 3, \quad mn \equiv 0 \ (3), \\ 0, & n \geqq 3, \quad mn \not\equiv 0 \ (3), \\ & \text{or } n = 1, 2 \text{ and } m = 1,2,3,\ldots \end{cases} .$$

Thus, $\kappa(L^{(n)}) = -\infty$, $n = 1, 2$, and $\kappa(L^{(n)}) = 0$ for $n \geqq 3$. Moreover, for $n \geqq 3$, $mn \equiv 0 \ (3)$, the effective m-th canonical divisor has the form

$$\sum_{i=1}^{3^n} m\left(\frac{n}{3} - 1\right)E_i,$$

where E_i appears in the canonical resolution of the singularities of $E_\rho^{(n)}/G$.

<u>Lemma 16.12.2</u>. $L^{(n)}$ is simply connected.

<u>Proof</u>. We use the fibration $f : L^{(n)} \longrightarrow L^{(1)} = \mathbb{P}^1$ induced by the projection

$$\begin{array}{ccc} E_\rho^n & \longrightarrow & E_\rho \\ \cup\!\!\!\!\!\; \shortmid & & \cup\!\!\!\!\!\; \shortmid \\ (z_1, \ldots, z_n) & \longmapsto & (z_1). \end{array}$$

Consider points $q_1 = [0]$, $q_2 = [\frac{1}{3} + \frac{2}{3}\rho]$, $q_3 = [\frac{2}{3} + \frac{1}{3}\rho]$ on $L^{(1)}$. The morphism f is of maximal rank at any point of $f^{-1}(\Delta')$, $\Delta' = L^{(1)} - \{q_1, q_2, q_3\}$ and $f : L^{(n)} \longrightarrow L^{(1)}$ is a fibre space of principally polarized abelian varieties (see Ueno [1], I, §1). It is easy to generalized the argument in Ueno [1], I, p.86~87 in our situation and we conclude that the singular fibres over three points q_1, q_2, q_3 on $L^{(1)}$ have the form

$$2L^{(n-1)} + \sum_{i=1}^{3^{n-1}} C_i ,$$

where C_i is one of the exceptional divisors E_k, $k = 1, 2, \ldots, 3^n$ appearing in Lemma 16.12.1. Note that C_i and $L^{(n-1)}$ intersect transversally.

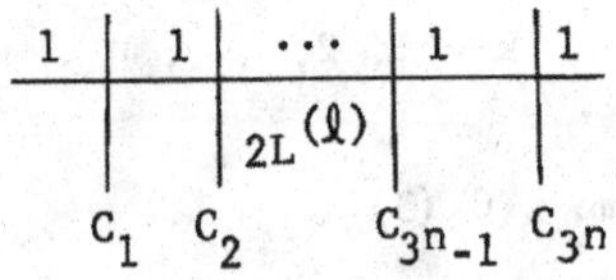

Now we shall prove the lemma by induction on n. As $L^{(1)}$ is $\mathbb{P}^1$, $L^{(1)}$ is simply connected.

Assume that $L^{(n-1)}$ is simply connected. Then the singular fibres $F_j = f^{-1}(q_j)$ are simply connected. As F_j is a compact algebraic set, F_j has a tubular neighbourhood V_j in $L^{(n)}$, which has a retraction to F_j. Hence V_j is simply connected. We set $L' = L^{(n)} - \bigcup_{j=1}^{3} F_j$, $V = \bigcup_{j=1}^{3} V_j$. Then L' is a torus bundle over Δ'.

$\pi_1(\Delta')$ is a free group generated by the homotopy class of loops τ_1, τ_2 which are small circles around q_1 and q_2, respectively, in Δ'. Let $\gamma_1, \gamma_2, \ldots, \gamma_{2(n-1)}$, be generators of the fundamental group of a fibre $f^{-1}(u)$ for $u \in \Delta'$. Then $\pi_1(L')$ is generated by $\gamma_1, \gamma_2, \ldots, \gamma_{2(n-1)}$, δ_1, δ_2, where δ_1 and δ_2 lie over τ_1 and τ_2, respectively. We can choose τ_i so small that there exists a neighbourhood W_j of q_j in $L^{(1)}$ where $\tau_j \subset W_j$ and $V_j \supset f^{-1}(W_j)$. As a consequence of Seifert's and Van Kampen's theorem, $\pi_1(L^{(n)})$ is trivial.

Q.E.D.

<u>Corollary 16.12.3</u>. If $n > 3$, $n \not\equiv 0$ (3), then the algebraic manifold $L^{(n)}$ has not a birational model V such that $3K_V$ is analytically trivial.

This is an easy consequence of Lemma 6.1.4.

<u>Lemma 16.12.4</u>.

$$H^1(L^{(n)}, \Theta) = 0 \quad \text{for} \quad n \geq 3.$$

This is proved in Ueno [3], I, § 8.

<u>Example 16.13</u>. E^n_ρ is the same as above. Let G be the cyclic group of order three of analytic automorphisms of E^n generated by an automotphism

$$g : E_\rho^n \longrightarrow E_\rho^n$$
$$\quad\ \ \cup \qquad\qquad\quad \cup$$
$$(z_1, z_2, \ldots, z_n) \longmapsto (\rho z_1\ \rho^2 z_2, \ldots, \rho^2 z_n) .$$

Then g has 3^n fixed points and the quotient space E_ρ^n/G has 3^n singular points corresponding to the fixed points. The resolution of these singularities is obtained by generalizing the process provided in Ueno [1], I, p.56 ~ 58 and we obtain in this way a non-singular model $M^{(n)}$ of the quotient space E_ρ^n/G. Then we have the following results:

1) $M^{(n)}$ is simply connected.

2) $$g_k(M^{(n)}) = \begin{cases} \binom{n-1}{k}, & k \equiv 0\ (3), \\ 0\ , & k \equiv 1\ (3), \\ \binom{n-1}{k-1}, & k \equiv 2\ (3), \end{cases}$$

where $k = 1, 2, 3, \ldots, n$.

3) $$P_m(M^{(n)}) = \begin{cases} 1 & m(2n - 1) \equiv 0\ (3), \\ 0 & m(2n - 1) \not\equiv 0\ (3), \end{cases}$$

for $n \geqq 2$, $m = 1, 2, 3, \ldots$.

Hence, $\kappa(M^{(n)}) = 0$ for $n \geqq 2$. $M^{(2)}$ is a K 3 surface.

<u>Example 16.14</u>. Let $E = E_{\sqrt{-1}}$ be an elliptic curve with period matrix $(1, \sqrt{-1})$. We set $E^n_{\sqrt{-1}} = \underbrace{E \times \cdots \times E}_{n}$.

Let G be the cyclic group of order four of analytic automorphisms of $E^n_{\sqrt{-1}}$ generated by an automorphism

$$g : E^n_{\sqrt{-1}} \longrightarrow E^n_{\sqrt{-1}}$$
$$\quad\ \ \cup \qquad\qquad\quad \cup$$
$$(z_1, z_2, \ldots, z_n) \longmapsto (\sqrt{-1}z_1, -\sqrt{-1}z_2, \ldots, -\sqrt{-1}z_n).$$

The automorphism g has 2^n fixed points and the automorphism g^2 has $4^n - 2^n$ fixed points outside of the fixed points of g. The automorphism g operates on the set of the above $4^n - 2^n$ fixed points and each $\langle g\rangle$ orbit consists of two points. This implies that the quotient space $E^n_{\sqrt{-1}}/G$ has $2^n+2^{n-1}(2^n-1)$ singular points corresponding to the fixed points. The resolution of these singularities is obtained by generalizing the process to be seen in Ueno [1], I, p.56~58 and we obtain in this way a non-singular model $N^{(n)}$ of the quotient space E^n_ρ/G. Then we have the following results :

1) $N^{(n)}$ is simply connected.

2) $$g_k(N^{(n)}) = \begin{cases} \binom{n-1}{k}, & k \equiv 0 \ (4), \\ \binom{n-1}{k-1}, & k \equiv 2 \ (4), \\ 0, & \text{otherwise}, \end{cases}$$

where $k = 1, 2, \ldots, n$.

3) $$P_m(N^{(n)}) = \begin{cases} 1, & m(3n-2) \equiv 0 \ (4) \\ 0, & m(3n-2) \not\equiv 0 \ (4) \end{cases}$$

for $n \geqq 2$, $m = 1, 2, 3, \ldots$.

Hence $\kappa(M^{(n)}) = 0$ for $n \geqq 2$. $M^{(2)}$ is a K 3 surface.

Example 16.15. Let $E^n_{\sqrt{-1}}$ be the same as above and let G be the cyclic group of order four of analytic automorphisms of $E^n_{\sqrt{-1}}$ generated by an automorphism

$$\begin{array}{cccc} g : & E^n_{\sqrt{-1}} & \longrightarrow & E^n_{\sqrt{-1}} \\ & \cup & & \cup \\ & (z_1, z_2, \ldots, z_n) & \longmapsto & (\sqrt{-1}z_1, \sqrt{-1}z_2, \ldots, \sqrt{-1}z_n). \end{array}$$

A non-singular model $F^{(n)}$ of the quotient space obtained by the canonical resolution of its singularities (16.10) has the properties listed below :

1) $F^{(n)}$ is simply connected.

2) $$g_k(F^{(n)}) = \begin{cases} \binom{n}{k}, & k \equiv 0 \ (4), \\ 0, & k \not\equiv 0 \ (4), \end{cases}$$

where $k = 1, 2, \ldots, n$.

3) $$P_m(F^{(n)}) = \begin{cases} 1, & n \geq 4 \\ & mn \equiv 0 \ (4), \\ 0, & n \geq 4 \\ & mn \not\equiv 0 \ (4), \\ & \text{or } n \leq 3 . \end{cases}$$

Hence $\kappa(F^{(n)}) = 0$ for $n \geq 4$ and $\kappa(F^{(2)}) = \kappa(F^{(3)}) = -\infty$.

From 2) we see that $g_1(F^{(2)}) = g_2(F^{(3)}) = 0$. It is not known whether $F^{(3)}$ is unirational or not.

The following example is due to Igusa [1].

Example 16.16. Let Δ be a lattice in $\mathbb{C}$ such that $E = \mathbb{C}/\Delta$ is an elliptic curve. Let x_i, y_i, $i = 1, 2, 3$ be complex numbers such that $x_1, y_2, x_3 - y_3 \notin \Delta$, $2x_1, 2y_2, 2(x_3 - y_3) \in \Delta$. Let σ_i, $i = 1, 2$ be analytic automorphisms of $E \times E \times E$ defined as follows.

$$\sigma_1 : (z_1, z_2, z_3) \longmapsto (z_1 + x_1, -z_2 + x_2, -z_3 + x_3) .$$

$$\sigma_2 : (z_1, z_2, z_3) \longmapsto (-z_1 + y_1, z_2 + y_2, -z_3 + y_3) .$$

The automorphism σ_i has no fixed point. We set $V = E \times E \times E/H$, where H is the Klein four group of analytic automorphisms of $E \times E \times E$ generated by σ_1 and σ_2. Then V is an algebraic manifold and has the following numerical invariants :

1) $$g_k(V) = \begin{cases} 0, & k = 1, 2. \\ 1, & k = 3. \end{cases}$$

2) $P_m(V) = 1$, for $m \geq 1$.

3) $E \times E \times E$ is a finite unramified covering of V. Hence, $q^*(V) = 3$ (note that $q(V) = 0$, by 1).)

Now we shall prove the following proposition which is a generalization of Corollary 16.11.3 and Corollary 16.12.3.

Proposition 16.17. Let A be an ℓ-dimensional abelian variety.

1) If $n \geq 3$, there does not exist a bimeromorphically equivalent model V of $A \times K^{(n)}$ (see Example 16.12) such that mK_V is trivial for a positive integer m.

2) If $n \geq 4$, there does not exist a bimeromorphically equivalent model W of $A \times L^{(n)}$ such that mK_W is trivial for an integer m.

Proof. Since the proofs are similar, we shall only prove the first part. We use freely the notations in 16.10 and Example 16.11. Let $f : A \times K^{(n)} \longrightarrow V$ be a bimeromorphic mapping of $A \times K^{(n)}$ onto V. From Theorem 2.13 (Theorem 2.13 also holds for Moishezon manifolds), we have a modification $g : \tilde{V} \longrightarrow A \times K^{(n)}$ obtained by finite succession of monoidal transformations with non-singular centers such that $h = f \circ g : \tilde{V} \longrightarrow V$ is a modification. Let $\mathcal{E}$ be the exceptional divisor appearing in the modification g (that is, as a set, $\mathcal{E}$ consists of those points at which g is not locally biholomorphic). We let ω be a non-zero m-tuple $(n+\ell)$-form on $A \times K^{(n)}$. By Lemma 16.11.1, the m-th canonical divisor of $A \times K^{(n)}$ defined by the zeros of ω has a form

$$\sum_{i=1}^{2^{2n}} m\left(\frac{n}{2} - 1\right) A \times E_i .$$

Therefore, the pull back $g^*(\omega)$ has zeros on $g^{-1}(A \times E_i)$ and $\mathcal{E}$. By Lemma 6.3, there is an m-tuple $(n+\ell)$-form ω_1 on V such that $h^*(\omega_1) = g^*\omega$. If, by the modification h, one of the irreducible components of $\mathcal{E}$ and $g^{-1}(A \times E_i)$, say D, is mapped onto a subvariety of codimension one in V, then, by Zariski's Main Theorem 1.11, h is biholomorphic at any point of $D - D \cap L$, where L is a nowhere dense analytic subset of $\tilde{V}$ such that $D \not\subset L$. It follows that ω_1 has zero on $h(D)$. Hence mK_V is not trivial. Suppose that, by the modification h, every irreducible component of $\mathcal{E}$ and $g^{-1}(A \times E_i)$ is mapped onto a subvariety of codimension at least two in V. We note that the element $(dz_1)^2$ of $H^0(T, \underline{S}^2(\Omega_T^1))$ (where $\underline{S}^2(\Omega_T^1)$ is a symmetric tensor product of Ω_T^1) induces an element of $H^0(A \times K^{(n)} - \bigcup_{i=1}^{2^{2n}} A \times E_i, \underline{S}^2(\Omega^1))$. Therefore $f^*(dz_1)^2$ is an element of $H^0(\tilde{V} - \tilde{\mathcal{E}}, \underline{S}^2(\Omega^1))$ where $\tilde{\mathcal{E}} = (\bigcup_{i=1}^{2^{2n}} A \times E_i) \cup \mathcal{E}$. By our assumption, h induces an isomorphism between $\tilde{V} - \tilde{\mathcal{E}}$ and $h(\tilde{\mathcal{E}})$ is an analytic set of codimension at least two. As $\underline{S}^2(\Omega_V^1)$ is a locally free sheaf, by Hartogs's theorem, $(h^{-1})^*(g^*(dz_1)^2)$ can be extended to a non-zero element of $H^0(V, \underline{S}^2(\Omega_V^1))$. It follows that $(dz_1)^2$ defines an element of $H^0(A \times K^{(n)}, \underline{S}^2(\Omega^1))$ (see 19.1 below, and Ueno [3], I, Proposition 1.2). But, by using the canonical resolution of Q_2^n in 16.10, we see that $(dz_1)^2$ can not be extended to a holomorphic section of $\underline{S}^2(\Omega_{\tilde{M}}^1)$ on $\tilde{M}$ where $\tilde{M}$ is a non-singular model of a neighbourhood of the singular point p in Q_2^n. This is a contradiction. Q.E.D.

Example 16.18. Let $\tilde{R}$ be a non-singular curve with an involution ι which has at least one fixed point and the quotient $R = \tilde{R}/\langle\iota\rangle$ is a non-singular curve of genus $g \geq 2$. We consider a surface S of general type in $\mathbb{P}^3$ defined by the equation

$$z_0^{10} + z_1^{10} + z_2^{10} + z_3^{10} = 0,$$

where $(z_0 : z_1 : z_2 : z_3)$ is a system of homogeneous coordinates of $\mathbb{P}^3$. The surface S has an involution

$$g : (z_0 : z_1 : z_2 : z_3) \longmapsto (z_0 : -z_1 : -z_2 : z_3).$$

The involution g has twenty fixed points and a non-singular model $\hat{S}$ of the quotient space $S/\langle g\rangle$ is a surface of general type. Let h be an involution of $\tilde{R} \times S$ defined by

$$\begin{array}{cccc} h : & \tilde{R} \times S & \longrightarrow & \tilde{R} \times S \\ & \text{\rotatebox{90}{$\in$}} & & \text{\rotatebox{90}{$\in$}} \\ & (x, y) & \longmapsto & (\iota(x), g(y)). \end{array}$$

The quotient space $\tilde{R} \times S/\langle h\rangle$ has isolated singularities. Each singular point of $\tilde{R} \times S/\langle h\rangle$ has a neighbourhood in $\tilde{R} \times S/\langle h\rangle$ which is analytically isomorphic to a neighbourhood of the singular point p in Q_2^3. Let V be a non-singular model of the quotient space $\tilde{R} \times S/\langle h\rangle$ obtained by the canonical resolution of its singularities (16.10). By our construction, there is a generically surjective rational mapping of V onto $R \times \hat{S}$. Since $R \times \hat{S}$ is of hyperbolic type, so is the threefold V. By a similar argument as that in the proof of Proposition 16.17, we infer that, for any bimeromorphically equivalent model V^* of V, the pluricanonical system $|mK(V^*)|$ has always fixed components for any positive integer. For the detailed discussion, see Ueno [7].

§ 17. Complex parallelizable manifolds

Nakamura [1] has studied deformations of parallelizable manifolds of dimension three and has shown that g_k, $h^{p,q}$, P_m, κ, q, r, t are not deformation invariants. In this section we shall give an outline of his results. For the details, we refer the reader to Nakamura [1]. In what follows, we shall freely use the results on the theory of harmonic integrals. For these results, see Kodaira and Morrow [1].

Definition 17.1. A compact complex manifold M is called a complex parallelizable manifold if the tangent bundle T_M of M is analytically trivial.

From this definition, if M is parallelizable, we have

$$g_k(M) = \binom{n}{k}, \quad k = 1, 2, \ldots, n = \dim M,$$
$$P_m(M) = 1, \qquad m = 1, 2, 3, \ldots ,$$
$$\kappa(M) = 0.$$

The following theorem is due to Wang [1].

Theorem 17.2. Let M be a complex parallelizable manifold. Then there exist a simply connected, connected, complex Lie group G and a discrete subgroup Γ of G such that $M = G/\Gamma$.

Definition 17.3. A complex parallelizable manifold $M = G/\Gamma$ is called solvable (resp. nilpotent) if the Lie group G is solvable (resp. nilpotent).

For a parallelizable manifold M, in an obvious way, we can define a pairing :

$$H^0(M, \Omega^p_M) \times H^0(M, \overset{p}{\wedge}\Theta) \longrightarrow \mathbb{C}$$
$$\varphi \times \theta \longmapsto (\varphi, \theta).$$

Then the exterior differentiation $d : H^0(M, \Omega^p_M) \longrightarrow H^0(M, \Omega^{p+1}_M)$ induces an adjoint map ${}^t d : H^0(M, \overset{p+1}{\wedge}\Theta) \longrightarrow H^0(M, \overset{p}{\wedge}\Theta)$.

The following formulae can be easily shown :

$$(18.4)\begin{cases} {}^t d(\theta \wedge \theta') = -[\theta, \theta'], \quad \text{for } \theta, \theta' \in H^0(M, \Theta), \\ {}^t d(\theta\wedge \theta' \wedge \theta'') = -\theta \wedge {}^t d(\theta' \wedge \theta'') - \theta' \wedge {}^t d(\theta'' \wedge \theta) - \theta'' \wedge {}^t d(\theta \wedge \theta'), \end{cases}$$

for $\theta, \theta', \theta'' \in H^0(M, \Theta)$.

The structure of the complex Lie group G is completely determined by its Lie algebra $\mathfrak{g}$. If $M = G/\Gamma$, then $\mathfrak{g}$ is isomorphic to $H^0(M, \Theta)$ with its complex Lie algebra structure. Let $\{\theta_1, \theta_2, \ldots, \theta_n\}$ be a basis of $H^0(M, \Theta)$. We have

$$(18.5) \qquad [\theta_\lambda, \theta_\mu] = \sum c'_{\mu\lambda\nu}\, \theta_\mu .$$

The structure of the Lie algebra $\mathfrak{g}$ is completely determined by the structure constants $\{c'_{\mu\lambda\nu}\}$. Let $\{\varphi_1, \varphi_2, \ldots, \varphi_n\}$ be the dual basis of $H^0(M, \Omega^1_M)$ with respect to the above basis $\{\theta_1, \theta_2, \ldots, \theta_n\}$. Using the first formula of 18.4, we can express 18.5 by the dual basis via

$$d\varphi_\mu = -\sum_{\lambda > \nu} c'_{\mu\lambda\nu}\, \varphi_\lambda \wedge \varphi_\nu$$

$$= \sum_{\lambda, \nu} c_{\mu\lambda\nu}\, \varphi_\lambda \wedge \varphi_\nu ,$$

where

$$c_{\mu\lambda\nu} = -c_{\mu\nu\lambda} = -\frac{1}{2} c'_{\mu\lambda\nu} .$$

Now suppose that M is a solvable manifold. Then we can choose a basis $\{\theta_1, \theta_2, \ldots, \theta_n\}$ of $H^0(M, \Theta)$ is such a way that $c'_{\mu\lambda\nu} = 0$ for $\mu < \max(\lambda, \nu)$. Hence, for a dual basis $\{\varphi_1, \varphi_2, \ldots, \varphi_n\}$, we have

$$d\varphi_\mu = 2(\sum_\lambda c_{\mu\lambda\mu}\varphi_\lambda)\wedge\varphi_\mu + \sum_{\lambda,\nu<\mu} c_{\mu\lambda\nu}\varphi_\lambda\wedge\varphi_\nu \quad .$$

We set $A_\mu = \sum_\lambda c_{\mu\lambda\mu}\varphi_\lambda$. It is easy to show that

$$d(\varphi_1\wedge\cdots\wedge\check{\varphi}_k\wedge\cdots\wedge\varphi_n)$$
$$= (-1)^{k-1}(A_1+A_2+\cdots+A_{k-1}+A_k+\cdots+A_n)\wedge(\varphi_1\wedge\cdots\wedge\check{\varphi}_k\wedge\cdots\wedge\varphi_n)$$

On the other hand, on an n-dimensional compact complex manifold, any holomorphic (n-1)-form is d-closed. Hence we have

$$\sum_\mu c_{\mu k\mu} = 0. \quad k = 1, 2, \ldots, n .$$

This implies that $\sum_{\lambda\mu} c_{\mu\lambda\mu}\varphi_\lambda = 0$. Using this equality, Nakamura has classified complex solvable Lie groups of dimension $\leqq 5$. We state the results in dimension three.

<u>Proposition 17.6</u>. Suppose that M is a solvable manifold of dimension three. By an appropriate choice of a basis $\{\varphi_1, \varphi_2, \varphi_3\}$ of holomorphic 1-forms on M, the basis belongs to one of the following three classes :

1) $d\varphi_\lambda = 0$, $\lambda = 1, 2, 3$.
2) $d\varphi_1 = 0$,
   $d\varphi_2 = 0$,
   $d\varphi_3 = -\varphi_1\wedge\varphi_2$,
3) $d\varphi_1 = 0$,
   $d\varphi_2 = \varphi_1\wedge\varphi_2$,
   $d\varphi_3 = -\varphi_1\wedge\varphi_3$.

(17.7) The structures of the Lie groups G of dimension three appearing in Proposition 17.6 are given as follows. First, Nakamura has proved that any simply connected, connected complex solvable Lie group is analytically isomorphic to $\mathbb{C}^n$ as a complex manifold (not necessarily as a complex Lie group). There are three cases corresponding to the above three classes in Proposition 17.6.

Case 1). G is a complex vector group $\mathbb{C}^3$.

Case 2). Let 0 be the origin of $\mathbb{C}^3$. We set $\Phi_\nu(z) = \int_0^z \varphi_\nu$, $\nu = 1, 2$. As $\varphi_\nu, \nu = 1, 2$ are d-closed, $\Phi_\nu(z)$, $\nu = 1, 2$, are single valued holomorphic functions on $\mathbb{C}^3$. We have $\varphi_\nu = d\Phi_\nu$, $\nu = 1, 2$. Hence $d\varphi_3 = -d\Phi_1 \wedge d\Phi_2$, i.e., $d(\varphi_3 + \Phi_1 d\Phi_2) = 0$. We set $\Phi_3(z) = \int_0^z (\varphi_3 + \Phi_1 d\Phi_2)$. Then Φ_3 is a single valued holomorphic function on $\mathbb{C}^3$ and $\varphi_3 = d\Phi_3 - \Phi_1 d\Phi_2$. For $g \in \Gamma$, we set $z' = g(z)$. As φ_ν is Γ-invariant for $\nu = 1, 2$, $d\Phi_\nu(z') = d\Phi_\nu(z)$, $\nu = 1, 2$. Hence we have

$$\Phi_\nu(z') = \Phi_\nu(z) + \omega_\nu(g),$$

where $\omega_\nu(g)$ is a constant depending only on g. Since we have

$$\begin{aligned}\varphi_3(z') &= d\Phi_3(z') - \Phi_1(z')d\Phi_2(z') \\ &= d\Phi_3(z') - (\Phi_1(z) + \omega_1(g))\, d\Phi_2(z),\end{aligned}$$

we obtain

$$\Phi_3(z') = \Phi_3(z) + \omega_1(g)\,\Phi_2(z) + \omega_3(g),$$

where $\omega_3(g)$ is a constant depending only on g. Define a multiplication $*$ on $\mathbb{C}^3$ by

$$(z_1, z_2, z_3)*(y_1, y_2, y_3) = (z_1 + y_1,\ z_2 + y_2,\ z_3 + y_1 z_2 + y_3).$$

Under the multiplication $*$, $\mathbb{C}^3$ becomes a nilpotent complex Lie group G. Note that we can also write

$$G = \left\{ \begin{pmatrix} 1 & z_2 & z_3 \\ 0 & 1 & z_1 \\ 0 & 0 & 1 \end{pmatrix} \,\middle|\, z_i \in \mathbb{C} \right\},$$

in which case the multiplication is the usual matrix multiplication.

Case 3). We set $\Phi_1(z) = \int_0^z \varphi_1$, $\Phi_2(z) = \int_0^z e^{-\Phi_1}\varphi_2$ and $\Phi_3(z) = \int_0^z e^{\Phi_1}\varphi_3$. Φ_1, Φ_2 and Φ_3 are single valued holomorphic function on $\mathbb{C}^3$. From arguments similar to those above, we obtain

$$\Phi_1(z') = \Phi_1(z) + \omega_1(g) ,$$
$$\Phi_2(z') = e^{-\omega_1(g)}\Phi_2(z) + \omega_2(g)$$
$$\Phi_3(z') = e^{\omega_1(g)}\Phi_3(z) + \omega_3(g),$$

where $z' = g(z)$, $g \in \Gamma$ and $\omega_\lambda(g)$, $\lambda = 1, 2, 3$ are constants depending only on g. We define a multiplication $*$ on $\mathbb{C}^3$ by

$$(z_1, z_2, z_3)*(y_1, y_2, y_3) = (z_1+y_1, e^{-y_1}z_2+y_2, e^{y_1}z_3+y_3).$$

Then $(\mathbb{C}^3, *)$ is the desired solvable group G.

Example 17.8. An example of type 2). (Iwasawa manifold) We have already seen that the group of type 2) can be written in the form

$$G = \left\{ \begin{pmatrix} 1 & z_2 & z_3 \\ 0 & 1 & z_1 \\ 0 & 0 & 1 \end{pmatrix} \middle| \, z_i \in \mathbb{C} \right\} ,$$

with the usual matrix multiplication. Let Γ be a discrete subgroup of G defined by

$$\Gamma = \left\{ \begin{pmatrix} 1 & \omega_2 & \omega_3 \\ 0 & 1 & \omega_1 \\ 0 & 0 & 1 \end{pmatrix} \middle| \, \omega_i \in \mathbb{Z}[\sqrt{-1}] \right\} .$$

The quotient manifold $M = G/\Gamma$ is a complex parallelizable manifold of type 2). M is called the Iwasawa manifold.

A basis of $H^0(M, \Theta)$ and the dual basis of $H^0(M, \Omega^1_M)$ are given as follows ;

$$\theta_1 = \partial_1,\ \theta_2 = \partial_2 + z_1\partial_3,\ \ \theta_3 = \partial_3,$$

$$\varphi_1 = dz_1,\ \varphi_2 = dz_2,\ \ \varphi_3 = dz_3 - z_1 dz_2,$$

with $\partial_\lambda = \dfrac{\partial}{\partial z_\lambda}$, $\lambda = 1, 2, 3$.

The form $\Omega = \sqrt{-1}\sum_{\lambda=1}^{3} \varphi_\lambda \wedge \bar{\varphi}_\lambda$ defines a hermitian metric on M. Using the metric Ω, we can calculate the dual operator ϑ of $\bar\partial$ and a laplacian $\square = \bar\partial\vartheta + \vartheta\bar\partial$. It is easy to show that $H^1(M, \mathcal{O}_M)$ is spanned by harmonic forms $\bar\varphi_1$, $\bar\varphi_2$ and $H^1(M, \Theta)$ is spanned by harmonic forms $\theta_i\bar\varphi_\lambda$, $i = 1,2,3$, $\lambda = 1,2$ (see the proof of Theorem 17.10 below.). Thus we obtain the following :

<u>Lemma 17.8.1</u>.

$$h^{0,1}(M) = 2, \quad r(M) = 2, \quad b_1(M) = 4.$$

Moreover, the Albanese mapping $\alpha : M \longrightarrow A(M)$ is surjective and under α, M becomes an elliptic bundle over A(M).

Small deformations of Iwasawa manifold will be constructed as follows.

For vector valued (0,1)-forms ψ, τ, we define

$$[\psi, \tau] = \sum(\psi^\alpha \wedge \partial_\alpha \psi^\beta + \tau^\alpha \wedge \partial_\alpha \psi^\beta)\partial_\beta,$$

where $\psi = \sum \psi^\alpha \partial_\alpha$, $\tau = \sum \tau^\alpha \partial_\alpha$. By an easy calculation, we have

$$[\theta_i\bar\varphi_\lambda, \theta_k\bar\varphi_\nu] = [\theta_i, \theta_k]\bar\varphi_\lambda \wedge \bar\varphi_\nu .$$

We set a vector valued (0,1)-form $\varphi(t) = \sum \varphi_{k_1,\dots,k_6}(t_{11})^{k_1}\dots(t_{32})^{k_6}$ and solve the equation

$$\bar\partial\varphi(t) - \frac{1}{2}[\varphi(t), \varphi(t)] = 0$$

under the conditions

$$\varphi(0) = 0 ,$$

$$\left(\frac{\partial \varphi(t)}{\partial t_{i\lambda}}\right)_{t=0} = \theta_i \bar{\varphi}_\lambda .$$

One of the solution is

$$\varphi(t) = \sum_{i=1}^{3} \sum_{\lambda=1}^{3} t_{i\lambda} \theta_i \bar{\varphi}_\lambda - (t_{11}t_{22} - t_{21}t_{12})\theta_3 \bar{\varphi}_3 .$$

Hence, if $|t| = \sum |t_{i\lambda}| < \varepsilon$ for a sufficiently small ε, then $\varphi(t)$ determines a complex structure M_t and we obtain an analytic family of deformations $\{M_t\}$ of M depending on six effective parameters $t_{i\lambda}$. The complex structure M_t is given as follows. We solve the following system of differential equations.

$$\bar{\partial} \zeta_\nu - \varphi(t) \zeta_\nu = 0, \quad \nu = 1, 2, 3,$$

under the initial conditions

$$\zeta_\nu(0) = z_\nu , \ \nu = 1, 2, 3.$$

There are the solutions :

$$\text{(17.8.2)} \quad \begin{cases} \zeta_1 = z_1 + \sum_{\lambda=1}^{2} t_{1\lambda} \bar{z}_\lambda , \\ \zeta_2 = z_2 + \sum_{\lambda=1}^{2} t_{2\lambda} \bar{z}_\lambda , \\ \zeta_3 = z_3 + \sum_{\lambda=1}^{2} (t_{3\lambda} + t_{2\lambda} z_1) \bar{z}_\lambda + A(\bar{z}) - D(t) \bar{z}_3 , \end{cases}$$

where

$$A(\bar{z}) = \frac{1}{2}(t_{11}t_{21}\bar{z}_1^{\,2} + 2t_{11}t_{22}\bar{z}_1\bar{z}_2 + t_{12}t_{22}\bar{z}_2^{\,2}) ,$$

$$D(t) = t_{11}t_{22} - t_{21}t_{12}.$$

Then there exists a differentiable function $c(t)$ of $t_{i\lambda}$'s with $c(0) = 1$ such that

$$d\zeta_1 \wedge d\zeta_2 \wedge d\zeta_3 \wedge d\bar{\zeta}_1 \wedge d\bar{\zeta}_2 \wedge d\bar{\zeta}_3$$
$$= c(t) dz_1 \wedge dz_2 \wedge dz_3 \wedge d\bar{z}_1 \wedge d\bar{z}_2 \wedge d\bar{z}_3 .$$

Hence, if ε is sufficiently small, then for $|t| < \varepsilon$ the mapping

$\Phi : (z_1, z_2, z_3) \longrightarrow (\zeta_1, \zeta_2, \zeta_3)$ defined by (17.8.2) is a diffeomorphism between $\mathbb{C}^3$ and $M_t = \mathbb{C}^3/\Gamma_t$, where Γ_t is a group of anal tic automorphisms of $\mathbb{C}^3$ defined by

$$\zeta_1 \longmapsto \zeta_1 + \omega_1(t)$$
$$\zeta_2 \longmapsto \zeta_2 + \omega_2(t)$$
$$\zeta_3 \longmapsto \zeta_3 + \omega_3(t) + \omega_1 \zeta_2 + \left(\sum_{\lambda=1}^{2} t_{2\lambda}\bar{\omega}_\lambda\right)\zeta_1 + A(\bar{\omega}) - D(t)\bar{\omega}_3 ,$$

with $\omega_i(t) = \omega_i + t_{i1}\bar{\omega}_1 + t_{i2}\bar{\omega}_2$ for an element

$$\begin{pmatrix} 1 & \omega_2 & \omega_3 \\ 0 & 1 & \omega_1 \\ 0 & 0 & 0 \end{pmatrix} \in \Gamma .$$

Using the same method as in the proof of Theorem 17.10 below, we can calculate $h^{p,q}(M_t)$. For $|t| < \varepsilon$, with a sufficiently small ε, we obtain

$$h^{p,q}(M_t) = h^{p,q}(M), \quad r(M_t) = r(M),$$
$$P_m(M_t) = P_m(M) = 1, \text{ for } m \geqq 1 ,$$
$$\kappa(M_t) = \kappa(M) = 0 .$$

Example 17.9. An example of type 3).

Let A be a 2×2 unimodular matrix with $\operatorname{tr} A \geqq 3$. A has real eigenvalues α, α^{-1} $(\alpha > 1)$. There exists a non-singular real matrix P such that

$$\begin{pmatrix} \alpha & 0 \\ 0 & \alpha^{-1} \end{pmatrix} = P A P^{-1} .$$

Let T be a two-dimensional complex torus with period matrix $(P, \tau P)$ where $\tau \in \mathbb{C}$, $\operatorname{Im}(\tau) > 0$. Let Γ_1 be a group of analytic automorphisms of $\mathbb{C} \times T$ generated by automorphisms

$$g_1 : (z_1, z_2, z_3) \longmapsto (z_1 + 2\pi i, z_2, z_3)$$

$$g_2 : (z_1, z_2, z_3) \longmapsto (z_1 + \beta, \alpha z_2, \alpha^{-1} z_3),$$

with $\beta = \log \alpha > 0$. The quotient manifold $M = \mathbb{C} \times T/\Gamma_1$ is a complex parallelizable manifold of type 3). A basis of $H^0(M, \Theta)$ and the dual basis of $H^0(M, \Omega^1_M)$ are given by

$$\theta_1 = \partial_1, \ \theta_2 = e^{-z_1}\partial_2, \ \theta_3 = e^{z_1}\partial z_3.$$

$$\varphi_1 = dz_1, \ \varphi_2 = e^{z_1}dz_2, \ \varphi_3 = e^{-z_1}dz_3.$$

Using a hermitian metric $\sqrt{-1}\sum_{\lambda=1}^{3} \varphi_\lambda \wedge \bar{\varphi}_\lambda$ on M, we can see that $H^1(M, \underline{O}_M)$ is spanned by harmonic forms $\varphi_1^* = d\bar{z}_1, \ \varphi_2^* = e^{z_1}d\bar{z}_2, \ \varphi_3^* = e^{-z_1}d\bar{z}_3$.

Hence $H^1(M, \Theta)$ is spanned by $\theta_i\varphi_\lambda^*$, $i = 1, 2, 3$, $\lambda = 1, 2, 3$.

Lemma 17.9.1. $q(M) = 3$, $r(M) = 1$, $b_1(M) = 2$.

Moreover, the Albanese mapping $\alpha : M \longrightarrow A(M)$ is surjective and under α, M becomes a torus bundle over an elliptic curve $A(M)$.

Next we shall consider small deformations of M. First we shall construct a one parameter family of deformations of M. By means of our construction, M can be viewed as a quotient space $\mathbb{C}^* \times \mathbb{C}^2/\Gamma$ of $\mathbb{C}^* \times \mathbb{C}^2$ by the group Γ of analytic automorphisms of $\mathbb{C}^* \times \mathbb{C}^2$ generated by automorphisms

$$g : (w, z_2, z_3) \longmapsto (\alpha w, \alpha z_2, \alpha^{-1} z_3)$$

$$g_j : (w, z_2, z_3) \longmapsto (w, z_2 + \omega_{2j}, z_3 + \omega_{3j}),$$
$$j = 1, 2, 3, 4,$$

where $\begin{pmatrix} \omega_{21} & \omega_{22} & \omega_{23} & \omega_{24} \\ \omega_{31} & \omega_{32} & \omega_{33} & \omega_{34} \end{pmatrix} = (P, \ \tau P)$.

We set $W_t = \{(\zeta_1, \zeta_2, \zeta_3) \in \mathbb{C}^3 \mid \zeta_1 - t\bar{\zeta}_2 \neq 0\}$. Then $W_0 = \mathbb{C}^* \times \mathbb{C}^2$

and W_t are diffeomorphic by a mapping

$$\zeta_1 = w + t\bar{z}_2$$
$$\zeta_2 = z_2$$
$$\zeta_3 = z_3 .$$

Let Δ_t be the group of analytic automorphism of W_t generated by automorphisms

$$g : (\zeta_1, \zeta_2, \zeta_3) \longmapsto (\alpha\, \zeta_1, \alpha\, \zeta_2, \alpha^{-1} \zeta_3)$$
$$g_j : (\zeta_1, \zeta_2, \zeta_3) \longmapsto (\zeta_1 + t\bar{\omega}_{2j}, \zeta_2 + \omega_{2j}, \zeta_3 + \omega_{3j})$$
$$j = 1, 2, 3, 4,$$

where the ω_j's are the same as above. The quotient $M_t = W_t / \Delta_t$ is a compact complex manifold and $\{M_t\}_{|t|<\varepsilon}$ is a one parameter complex analytic family. By our construction $M_0 = M$.

Lemm 17.9.2. For $t \neq 0$, any holomorphic function on W_t can be extended to a holomorphic function on $\mathbb{C}^3$.

Proof. For $t \neq 0$, let φ_t be an analytic automorphism of $\mathbb{C}^3$ defined by

$$\varphi_t : (\zeta_1, \zeta_2, \zeta_3) \longmapsto (t\, \zeta_1, \zeta_2, \zeta_3).$$

Then W_t is analytically isomorphic to W_1 by φ_t.
Hence it is enough to consider the case $t = 1$. We let φ be an analytic automorphism of $\mathbb{C}^3$ defined by

$$\varphi : (\zeta_1, \zeta_2, \zeta_3) \longmapsto (\zeta_1 - \zeta_2, \sqrt{-1}(\zeta_1 + \zeta_2), \zeta_3).$$

The automorphism φ maps W_1 isomorphically onto an open set

$$W' = \{(\eta_1, \eta_2, \zeta_3) \mid \mathrm{Re}\ \eta_1 \neq 0 \text{ or } \mathrm{Re}\ \eta_2 \neq 0\}$$

in $\mathbb{C}^3$. The domain W' is a tube domain (i.e., for any real numbers c_1, c_2, c_3, if $(\eta_1, \eta_2, \zeta_3) \in W'$, then $(\eta_1 + \sqrt{-1}c_1, \eta_2 + \sqrt{-1}c_2, \eta_3 + \sqrt{-1}c_3) \in W'$).

On the other hand, any holomorphic function on a connected tube domain can be extended to a holomorphic function on the convex hull of this tube domain. (see Bochner and Martin [1], Chap. V. Theorem 9, p.92, Hörmander [1], Theorem 2.5.10.) In our case the convex hull of W' is $\mathbb{C}^3$. Q.E.D.

Corollary 17.9.3. For $t \neq 0$, the universal covering $\tilde{W}_t$ of W_t is not a Stein manifold. Hence, a fortiori, $\tilde{W}_t$ is not analytically isomorphic to $\mathbb{C}^3$.

Lemma 17.9.4.

$$P_m(M_t) = \begin{cases} 0, & t = 0 \\ -\infty, & t \neq 0, \quad m = 1, 2, \ldots\ldots \end{cases}$$

$$r(M_t) = \begin{cases} 1, & t = 0 \\ 0, & t \neq 0. \end{cases}$$

$$g_1(M_t) = \begin{cases} 3, & t = 0 \\ 0, & t \neq 0. \end{cases}$$

Proof. Suppose $t \neq 0$. An element $\varphi \in H^0(M_t, \underline{O}(mK(M_t)))$ can be expressed in the form

$$\varphi(\zeta_1, \zeta_2, \zeta_3)(d\zeta_1 \wedge d\zeta_2 \wedge d\zeta_3)^m,$$

where $\varphi(\zeta_1, \zeta_2, \zeta_3)$ is holomorphic on W_t and satisfies

$$(17.9.5.) \quad \begin{cases} \varphi(\zeta_1, \zeta_2, \zeta_3) = \alpha^m \varphi(\alpha\zeta_1, \alpha\zeta_2, \alpha^{-1}\zeta_3), \\ \varphi(\zeta_1, \zeta_2, \zeta_3) = \varphi(\zeta_1 + t\bar{\omega}_{2j}, \zeta_2 + \omega_{2j}, \zeta_3 + \omega_{3j}), j=1,2,3,4. \end{cases}$$

By Lemma 17.9.2, $\varphi(\zeta)$ is holomorphic on $\mathbb{C}^3$. By the first equality of 17.9.5, we have $\varphi(\zeta) = \zeta_3^m f(\zeta)$, where $f(\alpha\zeta_1, \alpha\zeta_2, \alpha^{-1}\zeta_3) = f(\zeta_1, \zeta_2, \zeta_3)$. From the second equality of 17.9.5 we infer that $f(\zeta) = 0$.

By similar methods, we can easily calculate $r(M_t)$ and $g_1(M_t)$.

Q.E.D.

(17.9.6) By a similar method as in the proof of Theorem 17.10, below, we can calculate $h^{p,q}(M_t)$. The following table is due to Nakamura [1].

M_t	g_1	r	Albanese dimension t	q	g_2	$h^{0,2}$	P_m	p_g	κ
t=0	3	1	1	3	3	3	1	1	0
t≠0	0	0	0	2	0	1	0	0	$-\infty$

Note that invariants g_k, $h^{p,q}$ are invariant under small deformations of Kähler manifolds (use the upper semi-continuity of $h^{p,q}$ under small deformations, and $b_\nu = \sum_{p+q=\nu} h^{p,q}$ for a Kähler manifold; small deformations of Kähler manifolds are Kähler; see Kodaira and Morrow [1], Theorem 4.6, p.180).

(17.9.7) The Kuranishi space of M is given as follows. We shall solve the differential equation

$$\bar\partial\varphi(t) - \frac{1}{2}[\varphi(t), \varphi(t)] = 0$$

under the conditions

$$\varphi(0) = 0,$$

$$\left(\frac{\partial\varphi(t)}{\partial t_{i\lambda}}\right) = \theta_i\bar\varphi_\lambda \quad , \quad i = 1, 2, 3, \quad \lambda = 1, 2, 3.$$

The solution is

$$\varphi(t) = \sum t_{i\lambda}\theta_i\bar\varphi_\lambda \quad ,$$

where $t_{i\lambda}$'s satisfy the following equations:

$$\text{(17.9.8)} \quad \begin{cases} t_{11}t_{13} = 0, \quad t_{11}t_{12} = 0, \quad t_{12}t_{13} = 0, \\ t_{21}t_{13}-2t_{11}t_{23}=0, \quad t_{12}t_{21}=0, \quad t_{12}t_{23}=0, \\ t_{31}t_{13} = 0, \quad 2t_{11}t_{32}-t_{31}t_{12}=0, \quad t_{13}t_{32} = 0. \end{cases}$$

Hence the Kuranishi space is given by the equations 17.9.8 in a small neighbourhood of the origin 0 of $\mathbb{C}^9$ with coordinates $t_{i\lambda}$, i=1,2,3,

$\lambda = 1,2,3$. The deformations are, in this case, highly obstructed and the Kuranishi space has many branches at the origin. The deformation which we first considered corresponds to the subspace in the Kuranishi space defined by the equations

$$t_{11}=0,\ t_{13}=0,\ t_{21}=0,\ t_{22}=0,\ t_{23}=0,\ t_{31}=0,\ t_{32}=0,\ t_{33}=0.$$

Finally, with respect to parallelizable manifolds of type 3), Nakamura has shown the following :

Proposition 17.9.8. Let M be a three dimensional solvable parallelizable manifold of type 3). Then either $q(M) = 3$ or 1.

An example of type 3) with $q(M) = 1$ can be found by slightly modifying the example above.

Now we shall give additional theorems on solvable parallelizable manifolds of arbitrary dimensions due to Nakamura.

Theorem 17.10. If M is a nilpotent parallelizable manifold, then $q(M) = r(M)$ and $b_1(M) = 2r(M)$.

Proof. We shall calculate $H^1(M, \underline{O}_M)$ by virture of the Dolbeault cohomology $H_{\bar\partial}^{0,1}(M)$ (see Kodaira and Morrow [1], Theorem 6.3, p.80)

We choose a basis $\{\theta_1, \ldots, \theta_n\}$ of $H^0(M, \Theta)$ such that

$$[\theta_\lambda, \theta_\mu] = \Sigma\, c'_{\mu\lambda\nu}\, \theta_\mu,$$

where $c'_{\mu\lambda\nu} = 0$ for $\mu \leqq \max(\lambda, \nu)$. Let $\{\varphi_1, \ldots, \varphi_n\}$ be the dual basis of $H^0(M, \Omega^1)$. We have

$$d\,\varphi_\mu = \sum_{\lambda\nu} c_{\mu\lambda\nu}\, \varphi_\lambda \wedge \varphi_\nu,$$

where

$$c_{\mu\lambda\nu} = -\frac{1}{2} c'_{\mu\lambda\nu} .$$

If φ is a differentiable (0,1)-form on M, we can write

$$\varphi = \sum_{\lambda=1}^{n} f_\lambda \bar{\varphi}_\lambda ,$$

where f_λ's are differentiable function on M. It follows that

$$\begin{aligned} \bar{\partial}\varphi &= \sum_{\lambda,\nu} (\bar{\theta}_\nu f_\lambda) \bar{\varphi}_\nu \wedge \bar{\varphi}_\lambda + \sum_{\lambda=1}^{n} f_\lambda d\bar{\varphi}_\lambda \\ &= \sum_{1 \leq \nu < \lambda \leq n} (\bar{\theta}_\nu f_\lambda - \bar{\theta}_\lambda f_\nu + 2\sum_{\mu=1}^{n} \bar{c}_{\mu\nu\lambda} f_\mu) \bar{\varphi}_\nu \wedge \bar{\varphi}_\lambda . \end{aligned}$$

For φ and $\psi = \sum_{\lambda=1}^{n} g_\lambda \bar{\varphi}_\lambda$, we define an inner product (φ, ψ) by

$$(\varphi, \psi) = \int_M \sum f_\lambda \bar{g}_\lambda d\, M,$$

where

$$d\, M = (\sqrt{-1})^{-n^2} \varphi_1 \wedge \cdots \wedge \varphi_n \wedge \bar{\varphi}_1 \wedge \ldots \wedge \bar{\varphi}_n .$$

Let g be a differentiable function on M. From Stokes' theorem, we obtain

$$(\varphi, \bar{\partial} g) = \int_M \sum f_\lambda \theta_\lambda \bar{g}\; d\, M = -\int_M (\sum \theta_\lambda f_\lambda) \bar{g}\; d\, M = (\vartheta\varphi, g),$$

where we set

$$\vartheta \varphi = -\sum_{\lambda=1}^{n} \theta_\lambda f_\lambda .$$

A differentiable (0,1)-form φ is called harmonic if $\bar{\partial}\varphi = 0$, and $\vartheta\varphi = 0$. In view of the theory of harmonic integrals, $H_{\bar{\partial}}^{0,1}(M)$ is isomorphic to the vector space of harmonic (0,1)-forms (see, for example, Kodaira and Morrow [1] Chap.3). Suppose φ is a harmonic (0,1)-form. Then

$$(17.11) \quad \begin{cases} \bar{\theta}_\nu f_\lambda - \bar{\theta}_\lambda f_\nu + 2\sum_{\mu=1}^{n} \bar{c}_{\mu\nu\lambda} f_\mu = 0, \\ \sum_{\lambda=1}^{n} \theta_\lambda f_\lambda = 0. \end{cases}$$

We define $\square = \vartheta\bar{\partial} + \bar{\partial}\vartheta = -\sum_{\lambda=1}^{n} \theta_\lambda \bar{\theta}_\lambda$. Then, for a differentiable function f, $\Box f = 0$ implies $\bar{\partial} f = 0$. This means that f is holomorphic, hence, a constant. By 17.11, we have

$$\Box f_\nu = -2 \sum_{\lambda,\mu=1}^{n} \lambda \bar{c}_{\mu\nu\lambda} \theta_\lambda f_\mu .$$

As M is nilpotent, $c_{\mu\nu\lambda} = 0$ if $\nu \geqq \mu$ or $\lambda \geqq \mu$. Thus, $\Box f_n = 0$. This implies that f_n is a constant. Then we have

$$\Box f_{n-1} = -2 \sum_{\lambda,\mu=1}^{n} \bar{c}_{\mu,n-1,\lambda} \theta_\lambda f_\mu$$

$$= -2 \sum_{\lambda=1}^{n} \bar{c}_{n,n-1,\lambda} \theta_\lambda f_n = 0.$$

Hence f_{n-1} must be a constant. In this way, by induction on ν, we infer that f_ν is a constant for any ν. Hence, if $\{\varphi_{\nu_1}, \ldots, \varphi_{\nu_r}\}$ is a basis of $H^0(M, d\underline{O}_M)$, $\{\bar{\varphi}_{\nu_1}, \ldots, \bar{\varphi}_{\nu_r}\}$ spann the vector space of harmonic (0,1)-forms. Therefore, $\dim H^1(M, \underline{O}_M) = r$.

Next we shall prove that $b_1 = 2r$. From an exact sequence

$$0 \longrightarrow \mathbb{C} \longrightarrow \underline{O}_M \xrightarrow{d} d\underline{O}_M \longrightarrow 0,$$

we have an exact sequence

$$0 \longrightarrow H^0(M, d\underline{O}_M) \longrightarrow H^1(M, \mathbb{C}) \longrightarrow H^1(M, \underline{O}_M) \to \cdots .$$

Hence we have

$$b_1(M) \leqq r(M) + q(M) = 2r(M).$$

On the other hand, it is easy to show that we have

$$H^1(M, \mathbb{C}) \supset H^0(M, d\underline{O}_M) \oplus \overline{H^0(M, d\underline{O}_M)} .$$

Hence, $b_1(M) \geqq 2r(M)$. Q.E.D.

Theorem 17.10 is generalized in the form below.

<u>Theorem 17.12</u>. Let $M = G/\Gamma$ be a solvable parallelizable manifold associated with a simply connected, connected solvable Lie group G and a discrete subgroup Γ of G. If the Lie algebra $\mathfrak{g}$

of G has a Chevalley decomposition (i.e., $\mathfrak{g} = \mathfrak{a} \oplus \mathfrak{n}$ where $\mathfrak{a}$ is a commutative subalgebra and $\mathfrak{n}$ is the maximal nilpotent ideal), then $b_1(M) = 2r(M)$.

We state another important result.

Theorem 17.13. Let $M = G/\Gamma$ and $\mathfrak{g}$ be the same as in Theorem 17.12. If $\mathfrak{g}$ has a Chevalley decomposition, then any small deformation M_t of M has $\mathbb{C}^n$ as its universal covering, where $n = \dim M$.

Remark 17.14. Let $M = G/\Gamma$ be a complex parallelizable manifold associated with a simply connected, connected complex Lie group G and a discrete subgroup Γ of G. If G is semi-simple and has no $SL(2, \mathbb{C})$ as a factor, then $b_1(M) = 0$, $g_1(M) = 0$, $r(M) = 0$ and any small deformation of M is rigid. Furthermore, there is a discrete subgroup Γ of $SL(2, \mathbb{C})$ with a compact quotient space. It is not known whether $SL(2, \mathbb{C})/\Gamma$ has non trivial deformations or not.

§18 Complex structures on a product of two odd-dimensional homotopy spheres

H. Hopf [1] has shown that $S^1 \times S^{2n-1}$ carries non-Kähler complex structures which are usually called Hopf manifolds. Later, Calabi and Eckmann [1] has introduced on $S^{2p+1} \times S^{2q+1}$ non-Kähler complex structures which are usually called Calabi-Eckmann manifolds. One-dimensional Hopf manifolds are nothing other than elliptic curves. Two-dimensional Hopf manifolds has been studied in detail by Kodaira [3], II, III. Recently Ma. Kato [2],[3] has studied complex structures on $S^1 \times S^5$ and obtained interesting results. On the other hand, Brieskorn and Van de Ven [1] have introduced complex structures on products of two homotopy sheres. Maeda [1] has also introduced new complex structures on $S^{2p+1} \times S^{2q+1}$. Morita [1] has classified topologically complex structures on products of S^1 and homotopy spheres. In this section, we shall give an outline of their results.

(18.1) The Hopf manifolds are defined as follows. Let G be the infinite cyclic group of analytic automorphisms of $\mathbb{C}^n - \{0\}$ generated by an automorphism

$$g : (z_1, z_2, \ldots, z_n) \longmapsto (\alpha z_1, \alpha z_2, \ldots, \alpha z_n),$$

where $|\alpha| \neq 1$. It is easy to show that G acts on $\mathbb{C}^n - \{0\}$ properly discontinuously and freely. The quotient manifold $H = \mathbb{C}^n - \{0\}/G$ is compact and is called a Hopf manifold. For a point $(z_1, z_2, \ldots, z_n)$ of $\mathbb{C}^n - \{0\}$, we write the corresponding point of H as $[z_1, z_2, \ldots, z_n]$. There exists a surjective morphism $\pi : H \longmapsto \mathbb{P}^{n-1}$ defined by

$$\pi : [z_1, z_2, \ldots, z_n] \longmapsto (z_1 : z_2 : \cdots : z_n) .$$

By this morphism π, H becomes an elliptic bundle over $\mathbb{P}^{n-1}$. The elliptic curve appeared as a fibre has fundamental periods

$\{1, \frac{1}{2\pi\sqrt{-1}} \log \alpha\}$.

It is easy to show that the Hopf manifold H is diffeomorphic to $S^1 \times S^{2n-1}$. The Hopf manifold H has the following numerical invariants.

Lemma 18.1.1. 1) $g_k(H) = 0, \quad k = 1,2,\ldots,n,$

$r(H) = 0.$

2) $q(H) = 1$. More generally we have

$h^{p,q} = 0, \quad p = 1,2,\ldots,n-1.$

$$h^{0,q} = \begin{cases} 1, & q = 0, 1, \\ 0, & q \geqq 2. \end{cases}$$

$$h^{n,q} = \begin{cases} 0, & q \leqq n-2 \\ 1, & q = n-1, n. \end{cases}$$

3) $a(H) = n - 1, \quad \kappa(H) = -\infty$

4) $\dim_{\mathbb{C}} H^0(H, \Theta) = n^2$ and H is a homogeneous manifold.

5) $\dim_{\mathbb{C}} H^1(M, \Theta) = n^2$

$H^k(M, \Theta) = 0, \quad k \geqq 2.$

The proof of 2) and 5) can be found in Ise [1]. Another part of the above lemma can be easily proved.

We can generalize the above construction of Hopf manifolds in the following way.

Let $(\alpha) = (\alpha_1, \alpha_2, \ldots, \alpha_n)$ be an n-tuple of complex numbers such that

$$0 < |\alpha_1| \leqq |\alpha_2| \leqq \cdots \leqq |\alpha_n| < 1 .$$

G is the infinite cyclic group of analytic automorphisms of $\mathbb{C}^n - \{0\}$ generated by an automorphism

$$g_\alpha : (z_1, z_2, \ldots, z_n) \longmapsto (\alpha_1 z_1, \alpha_2 z_2, \ldots, \alpha_n z_n).$$

The quotient space $H_{(\alpha)} = \mathbb{C}^n - \{0\}/G$ is a compact complex manifold. It is easy to show that the manifold $H_{(\alpha)}$ is a deformation of the above n-dimensional Hopf manifold H. Therefore, the manifold $H_{(\alpha)}$ is diffeomorphic to $S^1 \times S^{2n-1}$.

We shall consider the case where $\alpha_1 = \alpha_2 = \ldots = \alpha_m$, $2 \leqq m \leqq n-2$, $0 < |\alpha_{m+1}| < |\alpha_{m+2}| < \cdots < |\alpha_n| < 1$, and the complex numbers $\alpha_1, \alpha_{m+1}, \cdots, \alpha_n$ are algebraically independent over $\mathbb{Q}$. In this case, we have the following lemma. The proof is elementary and we leave it to the reader.

Lemma 18.1.2. Let $(\alpha) = (\alpha_1, \alpha_1, \ldots, \alpha_1, \alpha_{m+1}, \ldots, \alpha_n)$ be the same as above. Then we have

$$a(H_{(\alpha)}) = m - 1.$$

More precisely, any meromorphic function on $H_{(\alpha)}$ is induced from a meromorphic function $P(z)/Q(z)$ of $\mathbb{C}^n$ such that P and Q are homogeneous polynomials of m variables $z_1, z_2, \ldots, z_m$ of the same degree. Hence the meromorphic function field $\mathbb{C}(H_{(\alpha)})$ is a purely transcendental extension of m-1 variables over $\mathbb{C}$.

Corollary 18.1.3. Let S be a complex submanifold of codimension n-m of $H_{(\alpha)}$ defined by the equations

$$z_1 = 0, \; z_2 = 0, \; \ldots, \; z_m = 0,$$

and let $\pi : H^*_{(\alpha)} \longrightarrow H_{(\alpha)}$ be the monoidal transformations with center S. For a meromorphic mapping

$$\begin{array}{ccc} g : H_{(\alpha)} & \longmapsto & \mathbb{P}^{m-1} \\ \psi & & \psi \\ [z_1, z_2, \ldots, z_m] & \longmapsto & (z_1 : z_2 : \cdots : z_m) \end{array}$$

of $H_{(\alpha)}$ onto $\mathbb{P}^{m-1}$, the composition $\varphi = g \circ \pi : H^*_{(\alpha)} \longrightarrow \mathbb{P}^{m-1}$ is a morphism. Moreover, the morphism $\varphi : H^*_{(\alpha)} \longrightarrow \mathbb{P}^{m-1}$ is the algebraic reduction of $H_{(\alpha)}$.

Remark 18.1.4. It is easy to see that the above meromorphic mapping $g : H_{(\alpha)} \longrightarrow \mathbb{P}^{m-1}$ is not a morphism. Hence there does not exist the algebraic reduction of $H_{(\alpha)}$ which has the form $\varphi : H_{(\alpha)} \longrightarrow \mathbb{P}^{m-1}$ (see Remark 12.7).

(18.2) To generalize the above procedure, we need the results from defferential topology.

Let $(a) = (a_0, \ldots, a_n)$ be an $(n+1)$-tuple of positive integers and let $X(a) = X(a_0, \ldots, a_n)$ be the affine algebraic variety in $\mathbb{C}^{n+1}$ defined by the equation

$$z_0^{a_0} + z_1^{a_1} + \cdots + z_n^{a_n} = 0.$$

Let $\Sigma(a)$ be the intersection of the affine variety $X(a)$ and the sphere S^{2n+1} in $\mathbb{C}^{n+1}$ defined by the equation

$$|z_0|^2 + |z_1|^2 + \cdots + |z_n|^2 = 1.$$

Since $X(a)$ is non-singular outside the origin, and $X(a)$ and the sphere intersect transversally, $\Sigma(a)$ is a differentiable manifold. Brieskorn [1] has shown that, for $n \neq 2$, $\Sigma(a)$ is a homotopy sphere if and only if $a_i = 1$ for some i (in this case $X(a)$ is non-singular at the origin, hence $\Sigma(a)$ is the standard sphere S^{2n-1}), or $a_j \geq 2$ for any j and $\Delta(1) = \pm 1$ where $\Delta(t) = \prod(t - \omega_0 \omega_1 \cdots \omega_n)$ such that each ω_k ranges over all a_j-th roots of unity other than 1 (see also Milnor [1]).

Theorem 18.2.1. (Brieskorn) For $n \neq 2$ and every $(2n-1)$-dimensional homotopy sphere Σ^{2n-1} bounding a parallelizable manifold, there are infinitely many $(n+1)$-tuple $a = (a_0, a_1, \ldots, a_n)$ such that the differentiable manifold $\Sigma(a)$ constructed above is diffeomorphic to Σ^{2n-1}.

For the proof, see Brieskorn [1]. Note that for $n \geqq 5$, an n-dimensional homotopy sphere is homeomorphic to the standard sphere S^n.

Now we shall define a holomorphic action of $\mathbb{C}$ on $X(a)-\{0\}$ by

$$t : (z_0, z_1, \ldots, z_n) \longmapsto (e^{t/a_0} z_0, e^{t/a_1} z_1, \ldots, e^{t/a_n} z_n), \quad t \in \mathbb{C}.$$

$\mathbb{Z}$ operates on $X(a)-\{0\}$ as the natural subgroup of $\mathbb{C}$. Since $\mathbb{Z}$ acts on $X(a)-\{0\}$ properly discontinuously and freely, we obtain the quotient manifold $H(a) = X(a) - \{0\}/\mathbb{Z}$. By construction, $H(a)$ is diffeomorphic to $S^1 \times \Sigma(a)$. If $a=(1,1,\ldots,1)$, $H(a)$ is nothing but the classical Hopf manifold defined in 18.1.

By a theorem of Holmann [1], the quotient space $V(a) = X(a)-\{0\}/\mathbb{C}$ exists as a normal complex space and there exists a surjective morphism $\pi_a : H(a) \longrightarrow V(a)$. There is a nowhere dense analytic subset S in $V(a)$ such that for any point $p \in V(a)-S$, the fibre $\pi_a^{-1}(p)$ is isomorphic to a fixed elliptic curve T_a with fundamental periods $\{1, 2\pi\sqrt{-1}[a]\}$, where $[a]$ is the least common multiple of the integers $a_0, \ldots, a_n$. Furthermore, the fibre space $\pi_a : H(a) \longrightarrow V(a)$ has the structure of a holomorphic Seifert principal fibre space with T_a as the fibre and the structure group in the sense of Holmann [2]. Brieskorn and Van de Ven [1] have shown the following :

<u>Theorem 18.2.2</u>. Let $\Sigma(a)$ be a homotopy sphere with $a_i \geqq 2$, $i = 0, 1, \ldots, n$.

1) $V(a)$ is non-singular if and only if one of the following two conditions is satisfied.

① For all i,j with $i \neq j$, the greatest common divisor (a_i, a_j) of a_i and a_j is 1.

② n is odd and up to a permutation of indices, $(a) = (a_0, 2b_1, \ldots, 2b_n)$ with $(a_0, 2) = 1$, $(a_0, b_j) = 1$ and $(b_i, b_j) = 1$ for all

$i, j,\ i \neq j$.

2) If $V(a)$ is non-singular, then $V(a)$ is analytically isomorphic to $\mathbb{P}^{n-1}$. If $V(a)$ is singular, then $V(a)$ is a finite ramified covering of $\mathbb{P}^{n-1}$. Hence, in both cases, $a(H(a)) = n - 1$.

For the proof except for assertion that $a(H(a)) = n - 1$, see Brieskorn and Van de Ven [1]. Since $a(H(a)) \geq a(V(a)) = n - 1$ and if $a(M) = \dim M$, then $b_1(M)$ is even (see Corollary 9.3 and Remark 9.4), we have $a(H(a)) = n - 1$.

(18.3) The above examples of complex structures on $S^1 \times \Sigma^{2n-1}$ (Σ^{2n-1} is a homotopy sphere which bounds a parallelizable manifold) do not exhaust all the complex structures on $S^1 \times \Sigma^{2n-1}$. In case of surfaces, all the complex structures on $S^1 \times S^3$ have been determined by Kodaira [3], II, III and Kodaira [5].

In case of dimension three Ma. Kato [2], [3] has shown the following interesting results.

<u>Theorem 18.3.1</u>. Let M be a three-dimensional complex manifold with $b_1(M) = 1$, $b_2(M) = 0$, $a(M) = 2$. Suppose that there is the algebraic reduction

$$\varphi : M \longrightarrow S,$$

where S is a non-singular algebraic surface. Then there exists an infinite cyclic unramified covering manifold W of M such that $W \cup \{\text{one point}\}$ has a complex structure of an affine variety with an algebraic $\mathbb{C}^*$ action.

For the proof, see Ma. Kato [2], Theorem 20.

<u>Theorem 18.3.2</u>. Let M be the same as above. Suppose that we have a complex analytic family

$$\pi \; : \; \underline{M} \longrightarrow D_{\varepsilon} = \{(t_1, \ldots, t_n) \in \mathbb{C}^n \mid |t_i| < \varepsilon\},$$

of the threefold M such that $M = M_0 = \pi^{-1}(0)$. Then we have a complex analytic family

$$\varpi \; : \; \underline{W} \longrightarrow D_{\varepsilon}$$

of $W = W_0 = \varpi^{-1}(0)$ such that there is a commutative diagram

$$\begin{array}{ccc} \underline{W} & \xrightarrow{\;f\;} & \underline{M} \\ & \searrow^{\varpi} \quad {}^{\pi}\swarrow & \\ & D_{\varepsilon} & \end{array},$$

where f is an infinite cyclic unramified covering of $\underline{M}$. Moreover, there exist a constant ε_0, $0 < \varepsilon_0 < \varepsilon$, a complex space $\hat{\underline{W}}$ and a surjective morphism $\hat{\varpi} : \hat{\underline{W}} \longrightarrow D_{\varepsilon_0}$ with the following properties.

1) $\hat{\underline{W}} = \varpi^{-1}(D_{\varepsilon_0}) \cup S$ where S is a closed analytic subset of $\hat{\varpi}^{-1}(D_{\varepsilon_0})$;

2) $\hat{\varpi}\big|_{\varpi^{-1}(D_{\varepsilon_0})} = \varpi$;

3) for any point $t \in D_{\varepsilon_0}$, $\hat{\varpi}^{-1}(t) \cap S$ is one point and $\hat{\varpi}^{-1}(t)$ is an affine variety.

For the proof and more detailed discussions, see Ma. Kato [2], Theorem 23 and Ma. Kato [3].

Note that, if M is a threefold whose underlying topological manifold is homeomorphic to $S^1 \times S^5$, then the condition that $b_1(M) = 1$ and $b_2(M) = 0$ is automatically satisfied.

(18.4) A topological classification of complex structures on $S^1 \times \Sigma^{2n-1}$ has been done by S. Morita [1].

Definition 18.4.1. Two almost complex structures M_0 and M_1 on a compact differentiable manifold M are called to be almost complex deformation equivalent (we shall write a.c. deformation equivalent) if there exists a differentiable family $\pi : \underline{M} \longrightarrow I = [0, 1]$ of almost complex manifolds over the unit interval such that $M_0 = \pi^{-1}(0)$ and $M_1 = \pi^{-1}(1)$ (see Kodaira and Spencer [1], I, Definition 1.6, p.336). Two complex structures on the differentiable manifold M are called to be almost complex deformation equivalent if the underlying almost complex structures are almost complex deformation equivalent.

From our definition, we have an equivalence relation on the set of all complex structures on a compact differentiable manifold. Note that, if a complex manifold M_0 is a deformation of a complex manifold M_1, then M_0 and M_1 are a.c. deformation equivalent.

The main theorem of Morita [1] is the following.

Theorem 18.4.2. Let Σ^{2n-1} be a (2n-1)-dimensional homotopy sphere which bounds a parallelizable manifold where $n > 2$.

1) If n is even, there are infinitely many complex structures on $S^1 \times \Sigma^{2n-1}$ which are not a.c. deformation equivalent.

2) If $n \equiv 1 \mod 4$, there exist exactly $(n - 1)!/2$ a.c. deformation equivalence classes of complex structures on $S^1 \times \Sigma^{2n-1}$.

3) If $n \equiv 3 \mod 4$, there exist exactly one or $(n - 1)!/4$ a.c. deformation equivalence classes of complex structures on $S^1 \times \Sigma^{2n-1}$ according as $n = 3$ or $n \neq 3$, respectively.

For the proof we refer the reader to Morita [1].

From the above theorem, there is only one a.c. deformation equivalence class of complex structures on $S^1 \times S^5$. By 18.2, complex manifolds $M(d) = H((2,2,2,d))$, where $d \equiv 1 \mod 2$, are diffeomorphic to $S^1 \times S^5$ (note that any five-dimensional homotopy sphere is diffeomorphic to the standard sphere S^5). It is interesting to know whether $M(d_1)$ is a complex analytic deformation of $M(d_2)$ or not.

(18.5) Calabi and Eckmann [1] have defined complex structures on $S^{2p+1} \times S^{2q+1}$ in the following way. Let $(x_0, x_1, \ldots, x_p)$ and $(y_0, y_1, \ldots, y_q)$ be global coordinates of $\mathbb{C}^{p+1}$ and $\mathbb{C}^{q+1}$, respectively. We shall consider the spheres S^{2p+1} and S^{2q+1} in $\mathbb{C}^{p+1}$ and $\mathbb{C}^{q+1}$ defined by the equations

$$|x_0|^2 + |x_1|^2 + \cdots + |x_p|^2 = 1 ,$$

$$|y_0|^2 + |y_1|^2 + \cdots + |y_q|^2 = 1 ,$$

respectively. There is a natural differential mapping

$$\begin{array}{cccc} \pi : & S^{2p+1} \times S^{2q+1} & \longrightarrow & \mathbb{P}^p \times \mathbb{P}^q \\ & \omega & & \omega \\ & (x_0, \ldots, x_p) \times (y_0, \ldots, y_q) & \longmapsto & (x_0 : \cdots : x_p) \times (y_0 : \cdots : y_q) . \end{array}$$

We shall define the complex structure $M_{p,q,\tau}$, $\mathrm{Im}(\tau) > 0$, on $S^{2p+1} \times S^{2q+1}$ as follows.

Let $V_{\alpha,\beta}$ be an open set in $S^{2p+1} \times S^{2q+1}$ defined by

$$x_\alpha \cdot y_\beta \neq 0 .$$

We set

$$z_\alpha^j = \begin{cases} \dfrac{x_{j-1}}{x_\alpha}, & j = 1, \ldots, \alpha , \\ \dfrac{x_j}{x_\alpha}, & j = \alpha+1, \ldots, p, \end{cases}$$

$$w_\beta^j = \begin{cases} \dfrac{y_{j-1}}{y_\beta}, & j = 1, \ldots, \beta , \\ \dfrac{y_j}{y_\beta}, & j = \beta+1, \ldots, q . \end{cases}$$

Let E_τ be the elliptic curve with fundamental periods 1 and τ and let $t_{\alpha\beta}(x, y)$, $x = (x_0, x_1, \ldots, x_p)$, $y = (y_0, y_1, \ldots, y_q)$, be the point $[\frac{1}{2\pi i}(\log x_\alpha + \tau \log y_\alpha)]$ of the elliptic curve E_τ. It is easy to see that there is a diffeomorphism

$$\tau_{\alpha\beta} : V_{\alpha\beta} \longrightarrow \mathbb{C}^{p+q} \times E_\tau$$
$$(x_0,\ldots,x_p)\times(y_0,\ldots,y_q) \longmapsto (z^1_\alpha,\ldots,z^p_\alpha, w^1_\beta,\ldots,w^q_\beta, t_{\alpha\beta}).$$

These diffeomorphisms $\tau_{\alpha\beta}$, $\alpha = 0,1,\ldots,p$, $\beta = 0,1,\ldots,q$, define a complex structure $M_{p,q,\tau}$ on $S^{2p+1}\times S^{2q+1}$ which is called a Calabi-Eckmann manifold. Moreover, the above differentiable mapping π: $S^{2p+1}\times S^{2q+1} \longrightarrow \mathbb{P}^p\times\mathbb{P}^q$ induces a surjective holomorphic mapping $\pi : M_{p,q,\tau} \longrightarrow \mathbb{P}^p\times\mathbb{P}^q$. By our construction, the fibre space $\pi: M_{p,q,\tau} \longrightarrow \mathbb{P}^p\times\mathbb{P}^q$ is a principal fibre bundle over $\mathbb{P}^p\times\mathbb{P}^q$ whose fibre and structure group are the elliptic curve E_τ.

The Calabi-Eckmann manifold $M_{p,q,\tau}$ is a homogeneous non-Kähler manifold of algebraic dimension $p+q$. We leave the proof to the reader.

(18.6) Maeda [1] has introduced complex structures on $S^{2p+1}\times S^{2q+1}$ which are fibre bundles over a "$\mathbb{P}^q$-bundle over $\mathbb{P}^p$" whose fibres and structure groups are an elliptic curve E and Aut(E), respectively.

Let $(y_0,y_1,\ldots,y_p)$ and $(x_{00},\ldots,x_{p0},x_{01},\ldots,x_{pq})$ be global coordinates of $\mathbb{C}^{p+1}$ and $\mathbb{C}^{(p+1)(q+1)}$, respectively. For a $(p+1)$-tuple $a = (a_0,a_1,\ldots,a_p)$ of positive integers with $a_0 \leqq a_1 \leqq \cdots \leqq a_p$, we define a submanifold $X_{p,a}$ of $\mathbb{C}^{p+1}\times\mathbb{C}^{(p+1)(q+1)}$ by the equations

$$y_k^{a_i} x_{i\ell} - y_\ell^{a_i} x_{ik} = 0 ,$$
$$i = 0,1,\ldots,0 , \quad k = 0,1,\ldots,q , \quad \ell = 0,1,\ldots,q .$$

We let S^{2p+1} and $S^{2(p+1)(q+1)-1}$ be spheres in $\mathbb{C}^{p+1}$ and $\mathbb{C}^{(p+1)(q+1)}$, respectively, defined by the equations

$$|y_0|^2 + |y_1|^2 + \cdots + |y_p|^2 = 1 ;$$
$$|x_{00}|^2 + |x_{10}|^2 + \cdots + |x_{p,q}|^2 = 1 .$$

We introduce the structure of a Calabi-Eckmann manifold

$\pi : M_{p,(p+1)(q+1)-1,\tau} \longrightarrow \mathbb{P}^p \times \mathbb{P}^{(p+1)(q+1)-1}$ on $S^{2p+1} \times S^{2(p+1)(q+1)-1}$

in $\mathbb{C}^p \times \mathbb{C}^{(p+1)(q+1)}$. We set $\tilde{\Sigma}_{p,a,\tau} = X_{p,a} \cap M_{p,(p+1)(q+1)-1,\tau}$.

$\tilde{\Sigma}_{p,a,\tau}$ is a complex submanifold of $M_{p,(p+1)(q+1)-1}$. The image $\pi(\tilde{\Sigma}_{p,a,\tau}) = \Sigma_{p,a}$ is a $\mathbb{P}^q$- bundle over $\mathbb{P}^p$ by the projection $\varpi : \Sigma_{p,a} \longrightarrow \mathbb{P}^p$ of $\Sigma_{p,a}$ onto the first factor. It is easily seen that the bundle $\varpi : \Sigma_{p,a} \longrightarrow \mathbb{P}^p$ is the projective fibre space associated with the vector bundle $[a_0 H] \oplus [a_1 H] \oplus \cdots \oplus [a_g H]$ where H is a hyperplane of $\mathbb{P}^p$ (see 2.8).

$$\begin{array}{ccc}
\tilde{\Sigma}_{p,a,\tau} & \hookrightarrow & M_{p,(p+1)(q+1)-1,\tau} \\
\searrow & & \downarrow \pi \\
\Sigma_{p,a} & \hookrightarrow & \mathbb{P}^p \times \mathbb{P}^{(p+1)(q+1)-1} \\
& \searrow \varpi & \downarrow \\
& & \mathbb{P}^p
\end{array}$$

<u>Proposition 18.6.1</u>. 1) $\tilde{\Sigma}_{p,a,\tau}$ is diffeomorphic to $S^{2p+1} \times S^{2q+1}$.

2) $\tilde{\Sigma}_{p,a,\tau}$ and $\tilde{\Sigma}_{p,b,\tau}$ are analytically isomorphic if and only if $a_i = b_i$, $i = 0,\dots,p$.

3) There exists a surjective group homomorphism $h : \mathrm{Aut}^0(\tilde{\Sigma}_{p,a,\tau}) \longrightarrow PGL(p, \mathbb{C})$ of the identity component of the automorphism group $\mathrm{Aut}(\tilde{\Sigma}_{p,a,\tau})$ of $\tilde{\Sigma}_{p,a,\tau}$ onto $PGL(p, \mathbb{C})$.

4) $\dim_{\mathbb{C}} \mathrm{Aut}^0(\tilde{\Sigma}_{p,a,\tau}) = \sum_{a_j \geq a_i} \binom{p+a_j-a_i}{p} + (p+1)^2 - 1$.

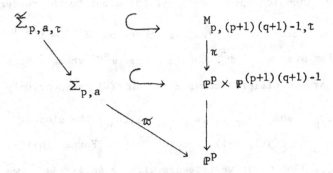

5) $\tilde{\Sigma}_{p,a,\tau}$ is a homogeneous manifold if and only if $a_0 = a_1 = \ldots = a_n$. In this case, $\tilde{\Sigma}_{p,a,\tau}$ is a Calabi-Eckmann manifold.

For the proof, see Maeda [1].

(18.7) Another complex structures on $S^{2p+1} \times S^{2q+1}$ have been constructed by Brieskorn and Van de Ven [1]. We use the same notation as in 18.2. Let $a = (a_0, a_1, \ldots, a_p)$ and $b = (b_0, b_1, \ldots, b_q)$ be tuples of natural numbers. For a complex number τ with $\mathrm{Im}(\tau) \neq 0$, we define an action f of $\mathbb{C}$ on $(X(a) - \{0\}) \times (X(b) - \{0\})$ by

$$f_\tau(t; x_0, \ldots, x_p; y_0, \ldots, y_q)$$
$$= (e^{t/a_0} x_0, \ldots, e^{t/a_p} x_p ; e^{\tau t/b_0} y_0, \ldots, e^{\tau t/b_q} y_q).$$

f_τ is a free, holomorphic locally proper action. Hence, by Holmann [1], Satz 10, the quotient manifold $H(a,b)_\tau = (X(a) - \{0\}) \times (X(b) - \{0\}) / \mathbb{C}$ exists. $H(a,b)_\tau$ is diffeomorphic to $\Sigma(a) \times \Sigma(b)$. On the other hand, a theorem of R. de Spaio (Notice of A.M.S., 15 (1968) p.628) says that a product of two homotopy spheres Σ and Σ' which bound parallelizable manifolds is always diffeomorphic to the product of the standard spheres. Note that the Calabi-Eckmann manifold $M_{p-1,q-1,-\tau}$ is isomorphic to the manifold $H(a;b)_\tau$ where $a = (1, 1, \ldots, 1)$, $b = (1, 1, \ldots, 1)$.

It would be interesting to consider deformations of the complex structures on $S^{2p+1} \times S^{2q-1}$ defined in 18.6, 18.7 and 18.8.

Chapter VIII

§ 19 Miscellaneous results

(19.1) Bimeromorphic invariants $g_k^m(V)$ and $\kappa^{(k)}(V)$.

For a complex variety V, Ueno [3] has defined

$$g_k^m(V), \quad k = 1,2,\dots, \dim V, \quad m = 1,2,\dots$$

by

$$g_k^m(V) = \dim_{\mathbb{C}} H^0(V^*, \underline{S}^m(\Omega_{V^*}^k)),$$

where V^* is a non-singular model of V and $\underline{S}^m(\Omega_{V^*}^k)$ is the m-th symmetric product of $\Omega_{V^*}^k$. The numbers $g_k^m(V)$ are well defined and are bimeromorphic invariants of the variety V. If dim V = n, $g_n^m(V)$ is nothing but the m-genus of the variety V. These numbers play an important role to study a ruled variety V. For example, if an n-dimensional variety V is bimeromorphically equivalent to $\mathbb{P}^1 \times W$, we have

$$g_{n-1}^m(V) = P_m(W),$$
$$g_k^m(V) = g_k^m(W), \quad k = 1, 2, \dots, n-2.$$

Let V be an n-dimensional complex manifold and let $M = \mathbb{P}(\Omega_V^k) \longrightarrow V$ be the projective fibre space associated with the locally free sheaf Ω_V^k. Since $\pi : M \longrightarrow V$ is a projective bundle over V, the spectral sequence

$$E_2^{p,q} = H^p(V, R^q\pi_*\underline{O}(mL)) \Longrightarrow H^{p+q}(M, \underline{O}(mL)),$$

degenerates for any positive integer m where L is the tautological line bundle of the projective bundle $\pi : M \longrightarrow V$ (see EGAII). Moreover, we have

$$\pi_*\underline{O}(mL) = \underline{S}^m(\Omega_V^k), \quad m = 1, 2, \dots ,$$

Hence, by Theorem 8.1, we have the inequalities

$$\alpha\, m^{\kappa^{(k)}(V)} \leqq g_k^{md}(V) \leqq \beta\, m^{\kappa^{(k)}(V)}$$

for suitable positive constants α, β and d where

$$\kappa^{(k)}(V) = \kappa(L, \mathbb{P}(\Omega_V^k)).$$

The numbers $\kappa^{(k)}(V)$, $k = 1,2,\ldots,n$ are bimeromorphic invariants of V. $\kappa^{(n)}(V)$ is nothing other than the Kodaira dimension of V. If V is bimeromorphically equivalent to the product $\mathbb{P}^1 \times W$, we have

$$\kappa^{(n-1)}(V) = \kappa(W).$$

(19.2) Birational invariants $Q_m(V)$ and $\nu(V)$.

Let V be an n-dimensional projective manifold and let D be a divisor on V. We write $D = \sum n_j D_j$ where the divisors D_j are prime divisors and $n_j \neq 0$. We define

$$P_m(D) = \sup_j P_m(D_j),$$

$$\kappa(D) = \sup_j \kappa(D_j).$$

Now we set

$$Q_m(D, V) = \inf\{ P_m(E) \mid E \sim rD,\ r \neq 0 \},$$

$$\nu(D, V) = \inf\{ \kappa(E) \mid E \sim rD,\ r \neq 0 \}.$$

Iitaka [4] has shown that $Q_m(K(V), V)$ and $\nu(K(V), V)$ are birational invariants. Hence, for an algebraic variety V, we define $Q_m(V)$ and $\nu(V)$ by

$$Q_m(V) = Q_m(K(V^*), V^*), \quad \nu(V) = \nu(K(V^*), V^*),$$

where V^* is a non-singular projective model of V.

Iitaka [4] has proved the following;

<u>Proposition 19.2.1</u>. Let $f : V \longrightarrow W$ be a fibre space of algebraic varieties where $\dim V = \dim W + 1$. There exists a Zariski

open set U of W such that, for any point $w \in W$, the fibre $V_w = f^{-1}(w)$ is irreducible and we have

$$\nu(V) \geqq \nu(V_w) + \kappa(W), \quad w \in W.$$

Proposition 19.2.2. Suppose that V is birationally equivalent to $\mathbb{P}^1 \times W$. Then we have

$$Q_m(V) = P_m(V),$$
$$\nu(V) = \kappa(W) .$$

For the proof, see Iitaka [4].

(19.3) Generalized adjunction formula.

Let S be a prime divisor on an algebraic manifold V. Iitaka [4] has shown

$$P_m(S) \leqq \ell(m(K(V) + S)|S),$$
$$\kappa(V) \leqq \kappa((K(V) + S)|S, S) ,$$

where for a Cartier divisor E on a variety W we have defined

$$\ell(m E) = \dim_{\mathbb{C}} H^0(W^*, \underline{O}(m \iota^* E)) ,$$

where $\iota : W^* \longrightarrow W$ is a normalization of W (see §8).

(19.4) Iitaka [5] has shown the following :

Theorem 19.4.1. Let V be a three-dimensional projective manifold. Suppose that the universal covering manifold of V is $\mathbb{C}^3$. Then the Kodaira dimension $\kappa(V)$ of V is neither 1 nor 3.

Iitaka has conjectured that, if V satisfies the above condition, there exists an abelian variety which is a finite unramified covering manifold of V. This is the case for curves and surfaces. For the detailed discussions, we refer the reader to Iitaka [5].

(19.5) Let W be an n-dimensional projective manifold and let $f : \mathbb{C}^n \longrightarrow W$ be a non-degenerate holomorphic mapping, i.e. the Jacobian of f does not vanish identically. For a prime divisor D on W, we denote by $\nu_x(f^*D)$ the multiplicity of the induced divisor f^*D at $x \in \mathbb{C}^n$. We say that f is ramified over D with multiplicity at least e, if $\nu_x(f^*D) \geq e$ holds at every point of the support of the divisor f^*D. If the image of f omits D, we say that f is ramified over D with multiplicity ∞.

Sakai [1] has shown the following :

Theorem 19.5.1. Let $D_1, \ldots, D_k$ be non-singular prime divisors on W such that $D = \sum_{i=1}^{k} D_i$ has only normal crossings. Let $f : \mathbb{C}^n \longrightarrow W$ be a non-degenerate holomorphic mapping which is ramified over D_i with multiplicity at least e_i for each i. Then

$$\kappa(K_W + \sum_{i=1}^{k} (1 - \frac{1}{e_i}) D_i, W) < n.$$

For the proof and the detailed discussions, see Sakai [1].

(19.6) A compact complex manifold V is called a prehomogeneous manifold if $\mathrm{Aut}(V)$ has an open orbit. This is equivalent to say that there exist n holomorphic vector fields $X_1, X_2, \ldots, X_n \in H^0(V, \Theta)$, $n = \dim X$ and a point $x \in V$ such that $X_1, X_2, \ldots, X_n$ induce a basis of the tangent space $T_x(V)$ of V at x. Hence, for a prehomogeneous manifold V, we have $\kappa(V) \leqq 0$.

Theorem 19.6.1. (Remmert and Van de Ven) Let V be a prehomogeneous manifold. The Albanese mapping $\alpha : V \longrightarrow A(V)$ induces a group homomorphism $\tilde{\alpha} : \mathrm{Aut}^0(V) \longrightarrow \mathrm{Aut}^0(A(V))$. Then we have the following :

1) α and $\tilde{\alpha}$ are surjective ;

2) every fibre of α is connected ;

3) $\alpha : V \longrightarrow A(V)$ has the structure of a fibre bundle whose fibre and structure group are $F = \alpha^{-1}(t)$ $t \in A(V)$ and Ker $\tilde{\alpha}$, respectively. Moreover, F is a prehomogeneous manifold.

For the proof, see Potters [1]. Using the above theorem, Akao has obtained the following results.

Corollary 19.6.2. A Kähler prehomogeneous manifold is an analytic fibre bundle over its Albanese torus $A(V)$ whose fibre F is a prehomogeneous manifold with $q(F) = 0$.

Proposition 19.6.3. Let V be a Kähler prehomogeneous manifold with $q(V) = 0$. V is a unirational projective manifold. Moreover if $\dim V \leqq 3$, V is rational.

Theorem 19.6.4. Let V be a Kähler prehomogeneous manifold.

1) If $q(V) = \dim V$, V is a complex torus.

2) Suppose that $q(V) = \dim V - 1$ or $\dim V - 2$. In case $q(V) = \dim V - 2$, we assume moreover that the fibre of the Albanese mapping of V is $\mathbb{P}^2$. Then there exists a flat vector bundle E of rank 2 or 3 on $A(V)$ such that V is isomorphic to the projective bundle $\mathbb{P}(E)$ associated with the vector bundle E. Conversely, every projective bundle on a complex torus associated with a flat vector bundle is prehomogeneous.

3) Suppose that $q(V) = \dim V - 2$. Then the prehomogeneous manifold V is obtained from a prehomogeneous $\mathbb{P}^2$-bundle over $A(V)$ by a finite number of equivariant blowing-ups and blowing-downs where by equivariant blowing-ups, we mean a blowing-up with a non-singular center in which the prehomogeneity is not violated.

For the proof and the detailed discussions, we refer the reader to Akao [1], [2].

(19.7) Fujita has begun the deep analysis of ample (not necessarily very ample) linear systems on algebraic varieties and obtained interesting results. Here we shall provide some of his results. For the proof and more detailed discussions, we refer the reader to Fujita [1], [2].

By a polarized variety (V, F), we mean a pair of an algebraic variety V and an ample (not necessarily very ample) line bundle (or ample divisor) on V. By definition, two polarized varieties (V, F) and (V', F') are isomorphic if there exists an isomorphism $f : V \longrightarrow V'$ such that $F = f^{*}F'$. For a polarized variety (V, F) of dimension n(i.e. dim V = n), we set

$$d(V, F) = c_1(F)^n[V] \in \mathbf{Z}^+$$

$$\Delta(V, F) = \dim V + d(V, F) - \dim H^0(V, \underline{O}_V(F)),$$

where $c_1(F)$ is the first Chern class of the line bundle F.

<u>Theorem 19.7.1</u>. Let (V, F) be a polarized variety and let L be a linear system contained in the complete linear system $|F|$. Then we have

$$\dim Bs(L) < \Delta(V, F) + \dim |F| - \dim L,$$

where Bs(L) is an algebraic subset of V consisting of the support of the fixed part and the base locus of the linear system L and dim Bs(L) = - 1 if Bs(L) is empty.

Using this theorem, Fujita has shown the following :

<u>Theorem 19.7.2</u>. Let (V, F) be a polarized manifold (i.e. V is non-singular) of dimension n. Suppose that $\Delta(V, F) = 0$.

Then the polarized manifold (V, F) is one of the following manifolds:

1) $(\mathbb{P}^n, H)$, where H is a hyperplane bundle of $\mathbb{P}^n$; in this case, $d(\mathbb{P}^n, H) = 1$;

2) (Q^n, H), where Q^n is a non-singular hyperquadric in $\mathbb{P}^{n+1}$ and H is a hyperplane bundle ; in this case $d(Q^n, H) = 2$;

3) $(\mathbb{P}(E), F)$, where $\mathbb{P}(E)$ is the projective fibre space over $\mathbb{P}^1$ associated with a direct sum of invertible sheaves of positive degree over $\mathbb{P}^1$ and F is the dual of the tautological line bundle; in this case, $d(\mathbb{P}(E), F) \geqq n$;

4) $(\mathbb{P}^2, 2H)$; in this case $d(\mathbb{P}^2, 2H) = 4$.

For a poralized manifold (V, F) of dimension n, we set

$$g(V, F) = 1 + \frac{1}{2}(K_V + (n - 1)F)\cdot F^{n-1} \quad .$$

The number $g(V, F)$ is always an integer.

Theorem 19.7.3. Suppose that, for a polarized manifold (V, F), $Bs(|F|)$ consists of at most finite number of points and that $g(M, F) \geqq \Delta(M, F)$.

1) If $d(M, F) \geqq 2\Delta(M, F) - 1$, then a general member of the complete linear system $|F|$ is non-singular.

2) If $d(M, F) \geqq 2\Delta(M, F)$, then $Bs(|F|) = \emptyset$.

3) If $d(M, F) \geqq 2\Delta(M, F) + 1$, then the line bundle F is very ample.

Using this theorem, Fujita has shown the following :

Theorem 19.7.4. Let (V, F) be a polarized manifold of dimension $n \geqq 2$. Suppose that $\Delta(V, F) = 1$. Then we have the following:

1) $g(V, F) \geqq 1$ and $d = d(V, F) \leqq 9$. Moreover, $d = 9$ if and only if $(V, F) \cong (\mathbb{P}^2, 3H)$.

2) Suppose that $\dim V \geqq 3$. $d = 8$ if and only if $(V, F) \simeq (\mathbb{P}^3, 2H)$. $d = 7$ if and only if $(V, F) \simeq (Q_p(\mathbb{P}^3), 2H-E_p)$ where $f : Q_p(\mathbb{P}^3) \longrightarrow \mathbb{P}^3$ is a monoidal transformation with center p (p is a point on $\mathbb{P}^3$), E_p is the exceptional divisor appearing in the monoidal transformation and H is a pull back of the plane line bundle on $\mathbb{P}^3$ by the morphism f.

3) $d = 4$ if and only if V is a non-singular complete intersection of type $(2, 2)$ in $\mathbb{P}^{n+2}$ and F is the hyperplane bundle.

4) $d = 3$ if and only if V is a non-singular hypercubic in $\mathbb{P}^{n+1}$ and F is the hyperplane bundle.

5) $d = 2$ if and only if V is a non-singular two sheeted branched covering of $\mathbb{P}^n$ whose branch locus in $\mathbb{P}^n$ is of degree $2g(V, F) + 2$, and F is the pull back of the hyperplane bundle on V.

Fujita has studied also the structures of polarized manifold (V, F) with $\Delta(V, F) = 1$, $d(V, F) = 1$ and $\Delta(V, F) = 2$.

(18.8) Horikawa [1] has studied deformations of holomorphic mappings. As an application of his deformation theory, Horikawa [2] has obtained deep results on deformations of quartic surfaces.

Mumford [6], Nakamura [2], Namikawa [1], [2] and Ueno [1] have studied degenerations of abelian varieties which are deeply related to the study of canonical bundle formulae for fibres spaces and Conjecture K_n. The almost homogeneous compactifications of algebraic torus due to Oda and Miyake [1], (for more general discussions, see Kempf et al [1]) are also related to the study of degenerations of abelian varieties.

Finally, we refer the reader to Nakano [1] and Fujiki and Nakano [1] concerning exceptional manifolds of the first kind.

Appendix

§20 Classification of surfaces

In this appendix, we shall give a brief outline of the classification theorey of algebraic and analytic surfaces due to Castelnuovo, Enriques and Kodaira. The detailed discussions can be found in Kodaira [2], [3] and Šafarevič [1]. In what follows, a surface is always assumed to be non-singular.

Definition 20.1. An irreducible curve C on a surface S is called an exceptional curve of the first kind if the curve C is isomorphic to $\mathbb{P}^1$ and the self-intersection number $C^2 = -1$.

From a theorem of Castelnuovo (see, for example, Grauert [2], Kodaira [2], II, Appendix), it follows that an exceptional curve C of the first kind is contractible, i.e. there exists a proper surjective morphism $f : S \longrightarrow \hat{S}$ of surfaces such that

1) $f(C) = p$ is a point and $\hat{S}$ is smooth at p ;
2) f induces an isomorphism between $S - C$ and $\hat{S} - p$;
3) $f : S \longrightarrow \hat{S}$ is a monoidal transformation of $\hat{S}$ with center p.

Let $f : S \longrightarrow \hat{S}$ be the same as above. We have

$$H_2(S, \mathbb{Z}) \simeq H_2(\hat{S}, \mathbb{Z}) \oplus \mathbb{Z}\cdot C.$$

Since $H_2(S, \mathbb{Z})$ is a finite generated $\mathbb{Z}$-module, we can always assume that a surface S does not contain any exceptional curve of the first kind.

Definition 20.2. A surface S is called a minimal model if any bimeromorphic mapping $f : \tilde{S} \longrightarrow S$ of a surface $\tilde{S}$ onto S is a morphism.

Theorem 20.3. (Castelnuovo, Enriques, Zariski, Kodaira) If the bimeromorphically equivalence class of an analytic surface has no minimal model, then S is a ruled surface, i.e. S is birationally equivalent to a product of $\mathbb{P}^1$ and a curve C.

For the proof, see Kodaira [3], III, Theorem 56. This theorem implies that if a surface S has no exceptional curves of the first kind and S is not a ruled surface, then S is a minimal model.

In what follows, we always assume that any surface S does not contain exceptional curves of the first kind.

Theorem 20.4. (Chow, Kodaira). If $a(S) = 2$, S is projective.

For the proof, see Chow and Kodaira [1].

Theorem 20.5. (Kodaira) If $a(S) = 1$, there exists a non-singular curve C and a surjective morphism $\varphi : S \longrightarrow C$ such that a general fibre is an elliptic curve.

For the proof, see Kodaira [2], I.

Definition 20.6. 1) A surface S is called a ruled surface of genus g if S is birationally equivalent to a product of $\mathbb{P}^1$ and a non-singular curve of genus g.

2) A surface S is called an elliptic surface if there is a surjective morphism $f : S \longrightarrow C$ of S onto a non-singular curve C such that a general fibre of f is an elliptic curve. A surface S is called an elliptic surface of general type if $\kappa(S) = 1$ (note that if $\kappa(S) = 1$, then S is an elliptic surface : see Theorem 6.11).

3) A surface S is called a K 3 surface if $q(S) = 0$ and K_S is analytically trivial.

4) A surface S is called an Enriques surface if $q(S) = 0$,

$P_2(S) = 1$ and $p_g(S) = 0$.

6) A surface S is called a hyperelliptic surface if $q(S) = 1$ and $12\,K_S$ is analytically trivial.

7) An algebraic surface of hyperbolic type (see Definition 11.1) is called an algebraic surface of general type.

8) A surface S belongs to class VII_0 if $b_1(S) = q(S) = 1$, $p_g(S) = 0$.

Now we shall give tables of classification of surfaces due to Kodaira. For the proof, see Kodaira [3].

Table 20.7.

Class	b_1	p_g	K	c_1^2	structure
I_0	even	0			algebraic
II_0	0	1	0	0	K 3 surface
III_0	4	1	0	0	complex torus
IV_0	even	+	$\neq 0$	0	elliptic surface of general type
V_0	even	+		+	algebraic surface of general type
VI_0	odd	+		0	elliptic surface
VII_0	1	0			?

Table 20.8.

b_1	P_{12}	P_2	K	C_1^2	structure
even	0	0			$\mathbb{P}^2$ or ruled
0	1	1	0	0	K 3 surface
4	1	1	0	0	complex torus
even	+		$\neq 0$	0	elliptic surface
even	+	+		+	algebraic surface of general type

odd	+				elliptic surface
1	0	0			?

Table 20.9. Classification of algebraic surfaces with $p_g = 0$.

c_1^2	q	P_{12}	P_2	structure
$c_1^2 < 0$	$\geqq 2$	0	0	ruled surface of genus q
$c_1^2 = 0$	1	0	0	ruled surface of genus 1
		+		elliptic surface
	0	+		elliptic surface
$9 \geqq c_1^2 > 0$	0	+	+	algebraic surface of general type
		0	0	$\mathbb{P}^2 (c_1^2=9)$, rational ruled $(c_1^2=8)$

The following table can be reduced from the above tables and the theory of elliptic surfaces due to Kodaira [2], [3].

Table 20.10. Classification of analytic surfaces of parabolic type.

b_1	q	P_{12}	p_g	structure
4	2	1	1	complex torus
3	2	1	1	elliptic surface with a trivial canonical bundle
2	1	1	0	hyperelliptic surface
1	1	1	0	elliptic surface of class VII_0 with a trivial mK for a positive integer $m > 1$
0	0	1	1	K 3 surface
		1	0	Enriques surface

Using the above classification tables, Iitaka [1], II has shown the following important theorem.

Theorem 20.11. The plurigenera and the Kodaira dimensions of surfaces are invariant under global deformations.

Now we shall give more detailed properties of each class of surfaces. We always assume that any surface has no exceptional curves of the first kind.

(20.12) Algebraic surfaces of general type.

A surface S of general type is characterized by

$$P_2(S) \geqq 1 , \quad c_1{}^2 > 0 .$$

The m-genus $P_m(S)$ of the surface S of general type is given by

$$P_m(S) = \frac{1}{2} m(m-1)c_1{}^2 + p_g - q + 1 ,$$

for $m \geqq 2$. The more important fact is the following.

<u>Theorem 20.12.1</u>. For $m \geqq 5$, the m-th canonical mapping $\Phi_{mK} : S \longrightarrow \mathbb{P}^N$ of a surface S of general type is a birational morphism. Moreover its image $S_m = \Phi_m(S)$ is a normal variety with only rational double points as possible singularities, if $m \geqq 5$.

For the proof and more detailed discussions see Kodaira [4] and Bombieri [1].

Using the above theorem , Popp [1], II has proved that the coarse moduli space of surfaces of general type exists as an algebraic space (see also Popp [2]).

(20.13) Elliptic surfaces.

Let $\pi : S \longrightarrow \Delta$ be an elliptic surface. There exist a finite number of points $p_1, p_2, \ldots, p_m$ on Δ such that $\pi|_{S'} : S' = \pi^{-1}(\Delta') \longrightarrow \Delta' = \Delta - \{p_1, p_2, \ldots, p_m\}$ is of maximal rank at any point of S'. The equation

$$f = p_i$$

defines a divisor C_i which we call the singular fibre of the elliptic surface S over the point p_i. The singular fibre C_i is called a multiple fibre if the greatest common divisor of $(n_1, \ldots, n_k)$ is strictly greater than 1 where

$$C_i = \sum_{i=1}^{k} n_i E_i$$

with E_i irreducible and $n_i > 0$. Singular fibres of elliptic surfaces are classified by Kodaira (see Kodaira [2], II). Moreover, in Kodaira [2], II, III and Kodaira [3], I, Kodaira has shown the way to construct all elliptic surfaces and studied their properties. For example, he has shown that the canonical bundle K of the elliptic surface $\pi : S \longrightarrow \Delta$ is given by

(20.13.1) $$K = \pi^*(\mathfrak{K}_\Delta - \mathfrak{f}) + \sum (m_\nu - 1)[P_\nu]$$

where $\mathfrak{K}_\Delta$ is the canonical bundle of the curve Δ, $\mathfrak{f}$ is the line bundle on Δ such that

$$\deg(\mathfrak{f}) = -(p_g(S) - q(S) + 1),$$

and $m_\nu P_\nu$'s are all multiple fibres of the elliptic surface $\pi : S \longrightarrow \Delta$. Kodaira has also shown that $\deg(\mathfrak{f}) \leqq 0$ and $\deg \mathfrak{f} = 0$ if and only if $\pi_{|S'} : S' \longrightarrow \Delta'$ is an analytic fibre bundle over Δ' whose fibre and structure group are an elliptic curve E and Aut(E), respectively. As a corollary of this formula 20.13.1, we have

$$P_m(S) = \dim | m(\mathfrak{K}_\Delta - \mathfrak{f}) + \sum [m(1 - \frac{1}{m_\nu})] a_\nu | + 1,$$

where the bracket [ ] denotes the Gauss symbol and a_ν is the point on Δ over which $m_\nu P_\nu$ is the multiple fibre of S. For example, if S is an elliptic surface of general type, we have

$$P_m(S) = m(2\pi - 1 - q + p_g) + \sum [m(1 - \frac{1}{m_\nu})] + 1 - \pi,$$

for $m \geqq 2$, where π is the genus of the curve Δ.

Using the canonical bundle formula 20.13.1, Iitaka [1],II has shown that, for an elliptic surface $\pi : S \longrightarrow \Delta$ of general type, Φ_{86mK} is a morphism for a positive integer m, and the fibre space $\Phi_{86mK} : S \longrightarrow \Phi_{86mK}(S)$ is isomorphic to $\pi : S \longrightarrow \Delta$ for a sufficiently large m. The fundamental theorem on classification theory (Theorem 6.11) can be considered as a generalization of this fact.

An elliptic surface $\pi : S \longrightarrow \Delta$ is called a basic elliptic surface if there is a holomorphic section $0 : \Delta \longrightarrow S$. All elliptic surfaces are obtained from basic elliptic surfaces by twisting the fibrations and logarithmic transformations (see Kodaira [2], II, §10, p.613-624 and Kodaira [3], I, p.766-771). The method to construct basic elliptic surfaces can be found in Kodaira [2], II, §8, p.578-603.

Recently, Miyaoka [1] has shown that any elliptic surface with the even first Betti number carries a Kähler metric.

(20.14) K 3 surfaces.

Any K 3 surface is a deformation of non-singular quartic surface in $\mathbb{P}^3$(see Kodaira [3], I, p.773-785). Therefore, any K 3 surface is simply connected. Kodaira [6] has constructed all analytic surfaces which have the same homotopy type as K 3 surfaces. It is not known whether these surfaces (we call them homotopy K 3 surfaces) are diffeomorphic to K 3 surfaces or not. The moduli of polarized K 3 surfaces has been studied by Pjateckiĭ-Šapiro and Šafarevič [1] but some parts of their arguments are not clear.

(20.15) Hyperelliptic surfaces.

Any hyperelliptic surface has a finite unramified covering manifold which is a product of two elliptic curves. Any hyperelliptic surface has two kinds of fibre spaces as elliptic surfaces. For more detailed discussions see Suwa [1].

(20.16) Enriques surfaces

A "general" Enriques surface is birationally equivalent to a surface in $\mathbb{P}^3$ defined by the equation

$$(z_1z_2z_3)^2 + (z_2z_3z_0)^2 + (z_3z_0z_1)^2 + (z_0z_1z_2)^2 + z_0z_1z_2z_3\,\psi(z_0,z_1,z_2,z_3) = 0 ,$$

where $(z_0:z_1:z_2:z_3)$ is a system of homogeneous coordinates of $\mathbb{P}^3$ and $\psi(z_0,z_1,z_2,z_3)$ is a general quadratic form. The fundamental group of an Enriques surface is $\mathbb{Z}/(2)$ and its universal covering manifold is a K 3 surface. Hence $2K_S$ is analytically trivial for any Enriques surface S. Moreover, any Enriques surface is an elliptic surface (see Enriques [1], Chap. VII, and Šafarevič [1] Chap. X).

(20.17) Rational and ruled surfaces.

Any rational surface free from exceptional curves of the first kind is either $\mathbb{P}^2$ or a $\mathbb{P}^1$– bundle over $\mathbb{P}^1$ (see Nagata [1] and Kodaira [3], IV, Theorem 49, p.1053). Any ruled surface of genus $g \geq 1$ free from exceptional curves of the first kind is a $\mathbb{P}^1$– bundle over a non-singular curve of genus g. For more detailed discussions see Nagata [1] and Maruyama [1].

(20.18) Surfaces of class VII_0.

By a Hopf surface, we mean an analytic surface whose universal covering manifold is $\mathbb{C}^2-\{(0, 0)\}$. Certain elliptic surfaces and Hopf surfaces belong to this class (see Kodaira [3], II). He has also shown that these surfaces are closed under deformations. The structure of Hopf surfaces and elliptic surfaces belonging to class VII_0 is deeply studied in Kodaira [3], II, III. For example, he has shown the following interesting results.

Theorem 20.18.1. If the second Betti number b_2 of a surface vanishes and if the fundamental group $\pi_1(S)$ of S contains an infinite cyclic subgroup of finite index, then S is a Hopf surface.

For the proof see Kodaira [3], III.

Every Hopf surface contains at least one curve. Conversely Kodaira has shown the following :

Theorem 20.18.2. Let S be a surface of algebraic dimension zero. Suppose that $b_1(S) = 1$, $b_2(S) = 0$ and S contains at least one curve. Then S is a Hopf surface.

For the proof see Kodaira [3], II.

Topological classification of Hopf surfaces has been done by Ma. Kato [1].

(20.19) Recently, Inoue has found new examples (we call them Inoue surfaces) of surfaces of class VII_0 (see Inoue [1]).

(20.19.1) Surfaces S_M.

Let $M \in SL(3, \mathbb{Z})$ be a unimodular matrix with one real and two non-real eigenvalues, $\alpha, \beta, \bar{\beta}$ where $\alpha\beta\bar{\beta} = 1$ and $\alpha > 1$. Let ${}^t(a_1, a_2, a_3)$ be a real eigenvector of α and ${}^t(b_1, b_2, b_3)$ an eigenvector of β. We let G_M be the group of analytic automorphisms of $H \times \mathbb{C}$ (H is the upper half plane) generated by the automorphisms

$$(w, z) \longmapsto (w + m_1a_1 + m_2a_2 + m_3a_3, z + m_1b_1 + m_2b_2 + m_3b_3),$$
$$(m_1, m_2, m_3) \in \mathbb{Z}^3,$$
$$(w, z) \longmapsto (\alpha w, \beta z).$$

G_M acts on $H \times \mathbb{C}$ properly discontinuously and freely. We set $S_M = H \times \mathbb{C}/G_M$. Then S_M is differentiably a 3-torus bundle over a circle, $b_1(S_M) = 1$ and $b_2(S_M) = 0$. Moreover, S_M has the following properties :

1) S_M contains no curves ;

2) $\dim_{\mathbb{C}} H^k(S_M, \Theta) = 0, \quad k = 0, 1, 2.$

(20.19.2) Surfaces $S^{(+)}_{N,p,q,r;t}$.

Let $N = (n_{kj}) \in SL(2, \mathbb{Z})$ be a unimodular matrix with two real eigenvalues α, $1/\alpha$ where $\alpha > 1$. We choose real eigenvectors ${}^t(a_1, a_2)$ and ${}^t(b_1, b_2)$ of N corresponding to α and $1/\alpha$, respectively. We fix integers $p, q, r (r \neq 0)$ and a complex number t. Let (c_1, c_2) be the solution of the following equation :

$$(c_1, c_2) = (c_1, c_2) \cdot {}^tN + (e_1, e_2) + \frac{b_1a_2 - b_2a_1}{r}(p, q)$$

where

$$e_1 = \frac{1}{2} n_{i_1}(n_{i_1} - 1)a_1b_1 + \frac{i}{2} n_{i_2}(n_{i_2} - 1)a_2b_2 + n_{i_1}n_{i_2}b_1a_2, \qquad i = 1,2.$$

Let $G^{+}_{N,p,q,r;t}$ be the group of analytic automorphisms of $H \times \mathbb{C}$ generated by automorphisms

$$g_0 : (w, z) \longmapsto (\alpha w, z+t)$$

$$g_i : (w, z) \longmapsto (w+a_i, z+b_iw+c_i), \quad i = 1, 2,$$

$$g_3 : (w, z) \longmapsto (w, z+(b_1a_2-b_2a_1)/r) .$$

$G^{(+)}_{N,p,q,r;t}$ acts on $H \times \mathbb{C}$ properly discontinuously and freely. The quotient manifold $S^{(+)}_{N,p,q,r;t} = H \times \mathbb{C}/G^{(+)}_{N,p,q,r;t}$ is differentiably a fibre bundle over a circle whose fibre is a circle bundle over a 2-torus. Moreover, the surface $G^{(+)}_{N,p,q,r;t}$ has $b_1 = 1$ and $b_2 = 0$. The surface has the following properties :

1) $S^{(+)}_{N,p,q,r;t}$ contains no curves ;

2) $\dim_{\mathbb{C}} H^0(S^{(+)}_{N,p,q,r;t}, \Theta) = \dim_{\mathbb{C}} H^1(S^{(+)}_{N,p,q,r;t}, \Theta) = 1$,
$\dim_{\mathbb{C}} H^2(S^{(+)}_{N,p,q,r;t}, \Theta) = 0$

3) $\mathcal{S}_{N,p,q,r} = \{S^{(+)}_{N,p,q,r;t}\}_{t\in\mathbb{C}}$ is a locally complete family of deformations.

(20.19.3) Surfaces $S^{(-)}_{N,p,q,r}$. Let $N = (n_{i_j}) \in GL(2, \mathbb{Z})$ be a matrix with $\det N = -1$ and real eigenvalues α, $-1/\alpha$ where $\alpha > 1$. We choose real eigenvectors ${}^t(a_1, a_2)$ and ${}^t(b_1, b_2)$ of N corresponding to α and $-1/\alpha$, respectively. We fix integers $p,q,r(r\neq 0)$ and define (c_1, c_2) to be a solution of the equation

$$-(c_1, c_2) = (c_1, c_2)\cdot {}^tN + (e_1, e_2) + \frac{b_1a_2 - b_2a_1}{r}(p, q),$$

where

$$e_i = \frac{1}{2}n_{i_1}(n_{i_1}-1)a_1b_1 + \frac{1}{2}n_{i_2}(n_{i_2}-1)a_2b_2 + n_{i_1}n_{i_2}b_1a_2, \quad i = 1, 2.$$

Let $G^{(-)}_{N,p,q,r}$ be the group of analytic automorphisms of $H\times\mathbb{C}$ generated by automorphisms

$$g_0 : (w, z) \longmapsto (\alpha w, -z)$$

$$g_i : (w, z) \longmapsto (w+a_i, z+b_iw + c_i), \quad i = 1, 2,$$

$$g_3 : (w, z) \longmapsto (w, z + (b_1a_2 - b_2a_1)/r).$$

$G^{(-)}_{N,p,q,r}$ operates on $H\times\mathbb{C}$ properly discontinuously and freely. The quotient surface $S^{(-)}_{N,p,q,r} = H\times\mathbb{C}/G^{(-)}_{N,p,q,r}$ has $S^{(+)}_{N^2,p_1,q_1,r;0}$ as its two-sheeted unramified covering surface for certain p_1 and q_1. Moreover, we can prove

$$\dim_{\mathbb{C}} H^k(S^{(-)}_{M,p,q,r}, \Theta) = 0, \quad k = 0,1,2.$$

Inoue has shown the following :

<u>Theorem 20.19.4</u>. Suppose that a surface S satisfies the following conditions :

1) $b_1(S) = 1$, $b_2(S) = 0$;

2) S contains no curves ;

3) there exists a line bundle F on S such that

$$\dim_{\mathbb{C}} H^0(S, \Omega^1 \otimes \underline{O}(F)) \neq 0 .$$

Then the surface is isomorphic to S_M, $S^{(+)}_{N,p,q,r:t}$ or $S^{(-)}_{N,p,q,r}$.

For the proof and more detailed discussions, see Inoue [1].

More recently Inoue has constructed new surfaces belonging to class VII_0 with $b_2 \neq 0$. The details will be found in Inoue [2].

BIBLIOGRAPHY

Abhyankar, S.

[1] Local Analytic Geometry, New York : Academic Press 1964.

Akao, K.

[1] On prehomogeneous compact Kähler manifolds (in Japanese). Master degree thesis, Univ. of Tokyo, (1972).

[2] On prehomogeneous compact Kähler manifolds. Proc. Int. Conf. on Manifolds and Related Topics in Topology. 365-372, Tokyo : Univ. of Tokyo Press 1974. (see also Proc. Japan Acad., 49 (1973), 483-485).

Andreotti, A.

[1] Recherches sur les surfaces irregulieres. Acad. Roy. Belg. Cl. Sci. Mém. Coll. 8° 4(1952).

[2] Recherches sur les surfaces algébriques irregulieres. Acad. Roy. Belg. Cl. Sci. Mém. Coll. 8° 7(1952).

Andreotti, A. and W. Stoll.

[1] Analytic and Algebraic Dependence of Meromorphic Functions, Lecture Note in Math. 234, Berlin-Heidelberg-New York : Springer 1971.

[2] Extension of holomorphic maps. Ann. of Math., 72 (1960), 312-349.

Artin, M.

[1] Algebraization of formal moduli, II. Existence of modification. Ann. of Math., 91 (1970), 88-135.

Artin, M. and D. Mumford.

[1] Some elementary examples of unirational varieties which are not rational. Proc. London Math. Soc. Ser.3 25 (1972),75-95.

Atiyah, M. F.

[1] Some examples of complex manifolds. Bonner Math. Schriften, 6 (1958).

[2] The signature of fibre-bundles. Global Analysis papers in honor of K. Kodaira.73-84, Univ. of Tokyo Press and Princeton Univ. Press, 1969.

Baily, W. L.

[1] On the moduli of Jacobian varieties. Ann. of Math., 71 (1960), 303-314.

Blanchard, A.

[1] Sur les variétés analytiques complexes. Ann. Sci. École Norm. Sup. 73 (1958). 157-202.

Bochner, S. and W. T. Martin.

[1] Several complex variables. Princeton,N.J.: Princeton Univ. Press, 1948.

Bombieri, E.

[1] Canonical models of surfaces of general type. Publ. Math. IHES, 42 (1973), 171-219.

Bourbaki, N.

[1] Topologie Générale, Chap.9. Paris ; Hermann, 1958.

Brieskorn, E.

[1] Beispiele zur Differentialtopologie von Singularitäten. Invent. Math., 2 (1966), 1-14.

Brieskorn, E. and A. Van de Ven.

[1] Some complex structures on products of homotopy spheres. Topology, 7 (1968), 389-393.

Calabi, E.

[1] On Kähler manifolds with vanishing canonical class. Algebraic geometry and topology, a symposium in honor of S. Lefschetz. (1957), 78-89, Princeton, N. J. : Princeton Univ. Press.

Calabi, E. and B. Eckmann.

[1] A class of compact complex manifolds. Ann. of Math. 58 (1953), 494-500.

Cartan, H.

[1] Quotient d'un espace analytique par un groupe d'automotphismes. Algebraic Geometry and Topology, Symposium in honor of S. Lefschetz, 90-102. Princeton, N.J.: Princeton Univ. Press, (1957).

[2] Quotient of complex analytic spaces. Contribution to function theory. 1-15, Bombay : Tata Inst.,1960.

Cartan, H. and S. Eilenberg.

[1] Homological Algebra. Princeton, N.J.: Princeton Univ. Press, 1956.

Castelnuovo, G.

[1] Memorie scelte. Bolgna : Nicola Zanichelli Editore, 1937.

Chow, W. L. and K. Kodaira.

[1] On analytic surfaces with two independent meromorphic functions. Proc. Nat. Acad. Sci. U.S.A., 38 (1952), 319-325.

Clemens, H. and P. Griffith.

[1] The intermediate Jacobian of the cubic threefold. Ann. of Math., 96 (1972), 281-356.

Conforto, F.

[1] Abelsche Funktionen und Algebraische Geometrie. Berlin-Heidelberg-New York : Springer 1956.

Curtis, C. W. and I. Reiner.

[1] Representation theory of finite groups and associative algebras. New York : Interscience 1962.

Deligne, P.

[1] Théorème de Lefschetz et critères de dégénérescence de suites spectrales, Publ. Math. IHES., 35 (1968), 107-126.

[2] Theorie de Hodge, II. Publ. Math. IHES., 40 (1971), 107-126.

Douady, A.

[1] Le problème des modules pour les sous-espaces analytiques compacts d'un espace analytiques donné. Ann. Inst. Fourier, 16 (1966), 1-98.

Enriques, F.

[1] Le Superficie Algebriche. Bologna : Nicola Zanichelli Editore, 1949.

[2] Memorie scelte di geometria, 3 volumes. Bologna : Nicola Zanichelli Editore, 1956, 1959, 1966.

Fáry, I.

[1] Cohomologie des variétés algébriques. Ann. of Math., 65 (1957), 21-73.

Fischer, W. and H. Grauert.

[1] Lokal-triviale Familien kompakter komplexen Mannigfaltigkeiten. Nachr. Akad. Wiss. Göttingen Math-Phys. KI, II, (1965), 89-94.

Forster, O. and K. Knorr.

[1] Ein Beweis des Grauertschen Bildgarbensatzes nach Ideen von B. Malgrange. Manuscripta Math., 5 (1971), 19-44.

Freitag, E.

[1] Über die Struktur der Funktionenkörper zu hyperabelschen Gruppen, I. J. Rein Angew. Math., 247 (1971), 97-117.

Frisch, J.

[1] Points de platitude d'un morphism d'espaces analytiques. Invent. Math., 4 (1967), 118-138.

Fujiki, A. and S. Nakano.

[1] Supplement to "On the inverse of monoidal transformation". Publ. Research Inst. Math. Sci. Kyoto Univ., 7 (1972), 637-644.

Fujita, T.

[1] On the Δ-genera of polarized varieties (in Japanese). Master degree thesis, Univ. of Tokyo, 1974.

[2] On the Δ-genera of polarized varieties. To appear (see also Proc. Japan Acad., 49 (1974), 800-802, ibid. 50 (1974), 173-174.)

Grauert, H.

[1] Ein Theorem der analytischen Garbentheorie und die Modulräume komplexer Strukturen. Publ. Math. IHES., 5 (1960).

[2] Über Modifikationen und exzeptionelle analytischen Mengen. Math. Ann., 146 (1962), 331-368.

Grauert, H. and R. Remmert

[1] Komplexe Räume. Math. Ann., 136 (1958), 245-318.

Grothendieck, A.

[1] Fondaments de géométrie algébrique. Paris, 1962.

[SGA I] Revètements Etales et Group Fondamental (SGA 1). Lecture Note in Math., 224 Berlin-Heidelberg-New York : Springer 1971.

Grothendieck, A. and J. Dieudonné.

[EGA] Eléments des géométrie algébrique, I, II, III_1, III_2. Publ. Math. IHES., 4 (1960), ibid. 8 (1961), ibid. 11 (1961), ibid. 17 (1963).

[EGA I_a] Éléments des géométrie algébriques, I. Berlin-Heidelberg-New York : Springer 1971.

Gunning, R. C.

[1] Lectures on Riemann Surfaces. Princeton, N. J. : Princeton Univ. Press 1965.

Gunning, R. C. and H. Rossi.

[1] Analytic Functions of Several Complex Variables. Englewood Cliffs. N. J. : Princeton Hall Inc. 1965.

Hartshorn, R.

[1] Ample Subvarieties of Algebraic Varieties. Lecture Note in Math., 156, Berlin-Heidelberg-New York : Springer 1970.

[2] Ample vector bundles on curves. Nagoya Math. J., 43 (1971), 73-89.

Hironaka, H.

[1] Resolution of singularities of an algebraic variety over a field of characteristic zero, I, II. Ann. of Math., 79 (1964), 109-326.

[2] Bimeromorphic smoothing of a complex analytic space. Preprint, Univ. of Warwick. (1971).

[3] Review of S. Kawai's paper. Math. Review, 32 (1966), 87-88.

[4] Flattening of complex analytic maps. Proc. Int. Conf. on Manifolds and Related Topics in Topology 313-322. Tokyo : Univ. of Tokyo Press 1974.

Hironaka, H. and H. Rossi.

[1] On the equivalence of inbeddings of exceptional complex spaces. Math. Ann., 156 (1965), 313-333.

Hirzebruch, F.

[1] Topological Methods in Algebraic Geometry (third edition). Berlin-Heidelberg-New York : Springer 1960.

[2] Hilbert modular surfaces. L'Ens. Math., 19 (1973), 183-281.

Hirzebruch, H. and A. Van de Ven.

[1] Hilbert modular surfaces and the classification of algebraic surfaces. Invent. Math., 23 (1974), 1-29.

Hitotumatu, S.

[1] Theory of Analytic Functions of Several Variables (in Japanese). Tokyo : Baihukan 1960.

Hodge, W. V. D.

[1] The theory and appications of harmonic integrals (second ed.).

Cambridge: Cambridge Univ. Press. 1952.

Hodge. W. V. D. and D. Pedoe.

[1] Methods of Algebraic Geometry, II. Cambridge: Cambridge Univ. Press 1952

Holmann H.

[1] Quotientenräume komplexer Mannigfatigkeit nach komplexen Lieschen Automorphismengruppe. Math. Ann., 139 (1960), 383-402.

[2] Seifertsche Faserräume. Math. Ann., 157 (1964), 138-166.

Hopf, H.

[1] Zur Topologie der komplexen Mannigfaltigkeiten. Courant Anniversary Volume, 167-185. New York: Interscience Publ. Inc. 1948.

Horikawa, E.

[1] On deformations of holomorphic maps,I,II,III. J.Math.Soc.Japan, 25(1973),371-396,to appear in J.Math.Soc.Japan, to appear.

[2] On deformations of quintic surfaces. To appear in Invent. Math., (see also Proc. Japan Acad., 49 (1973), 377-379 and Proc. Int. Conf. on Manifolds and Related Topics in Topology. 383-388, Tokyo : Univ. of Tokyo Press 1974.)

Hörmander, L.

[1] An introduction to complex analysis in several variables. Princeton, N. J.-Tronto-New York-London : Van Nostrand, 1966.

Igusa, J.

[1] On the structure of a certain class of Kähler varieties. Amer. J. Math., 76 (1954), 669-678.

[2] Fibre systems of jacobian varieties, I, II and III. Amer. J. Math., 78 (1956), 171-199, 745-760, ibid. 81 (1959), 453-476.

[3] A desingularization problem in the theory of Siegel modular functions. Math. Ann., 168 (1967), 228-260.

Iitaka, S.

[1] Deformations of compact complex surfaces, I, II and III. Global Analysis, papers in honor of K. Kodaira, Princeton Univ. Press and Tokyo Univ. Press (1969) 267-272, J. Math. Soc. Japan, 22 (1970), 247-261, ibid. 23 (1971), 692-705.

[2] On D-dimensions of algebraic varieties. J. Math. Soc. Japan, 23 (1971), 356-373.

[3] Genera and classification of algebraic varieties, I (in Japanese). Sugaku, 24 (1972), 14-27.

[4] On some new birational invariants of algebraic varieties and their application to rationality of certain algebraic varieties of dimension 3. J. Math. Soc. Japan, 24 (1972), 384-396.

[5] On algebraic varieties whose universal covering manifolds are

complex affine 3-spaces, I. Number Theory, Algebraic Geometry and Commutative Algebra, in honor of Y. Akizuki (1973), 95-146, Tokyo : Kinokuniya.

[6] The problem of rationality of function fields (in Japanese). Surikaiseki Kenkyusho Kokyuroku, 68 (1969), 76-99.

Inoue, M.

[1] On surfaces of class VII_0. Invent. Math., 24(1974) 269-310.

[2] New surfaces with no meromorphic functions. To appear in Proc. Int. Congress Math., Vancouver.

Ise, M.

[1] On geometry of Hopf manifolds. Osaka Math.J., 12(1960),387-402.

Iskovski, V. A. and Ju. I. Manin.

[1] Three-dimensional quartics and counterexamples to the Lüroth Problem. Math. Sbonik 86 (128) (1971), English translation Math. USSR Sbornik 15 (1971), 141-166.

Kas, A.

[1] Deformations of elliptic surfaces. Thesis, Stanford Univ. 1966 (see also Proc. Nat. Acad. Sci. U.S.A., 58 (1967), 402-404).

[2] On deformation of a certain type of irregular algebraic surface. Amer. J. Math., 90 (1968), 789-804.

[3] Ordinary double points and obstructed surfaces.To appear.

Kato, Masahide.

[1] Topology of Hopf surfaces. To appear in J. Math. Soc. Japan.

[2] Complex structures on $S^1 \times S^5$. To appear in Invent. Math. (see also Proc. Japan Acad., 49 (1973), 575-577).

[3] A generalization of Bieberbach's example. Proc. Japan Acad., 50 (1974), 329-333.

Kato, Mitsuyoshi.

[1] Partial Poincaré duality for k-regular spaces and complex algebraic sets. To appear.

Kawai, S.

[1] On compact complex analytic manifold of complex dimension 3, I and II. J. Math. Soc. Japan, 17 (1965), 438-442, ibid., 21 (1969), 604-616.

[2] Elliptic fibre spaces over compact surfaces. Commen. Math. Univ. St. Paul., 15 (1967), 119-138.

Kempf, G. et al.

[1] Toroidal Embeddings I. Lecture Note in Math., 339, Berlin-Heidelberg-New York : Springer 1973.

Kiehl, R.

[1] Note zu der Arbeit von J. Frisch : "Points de platitude d'un morphism d'espaces analytiques". Invent. Math., 4 (1967), 139-141.

Kiehl, R . and J. L. Verdier.

[1] Ein einfacher Beweis des Koharenzsatzes von Grauert. Math. Ann., 195 (1972) 24-50.

Knorr, K.

[1] Der Grauertsche Projektionssatz. Invent. Math., 12 (1971), 118-172.

Kobayashi, S.

[1] Hyperbolic manifolds and holomorphic mappings. New York : Marcel Dekker 1970.

[2] Transformation groups in differential geometry. Berlin-Heidelberg-New York : Springer 1972.

Kodaira, K.

[1] On Kähler varieties of restricted type (an intrinsic characterization of algebraic varieties). Ann. of Math., 60 (1954), 28-48.

[2] On compact analytic surfaces, I, II and III. Ann. Math., 71 (1960), 111-152, ibid. 77 (1963), 563-626, ibid. 78 (1963), 1-40.

[3] On the structure of compact complex analytic surfaces, I, II, III and IV. Amer. J. Math., 86 (1964), 751-798, ibid. 88 (1966), 682-721, ibid. 90 (1968) 55-83, 1048-1066.

[4] Pluricanonical systems of algebraic surfaces of general type. J. Math. Soc. Japan, 20 (1968), 170-192.

[5] Complex structures on $S^1 \times S^3$. Proc. Nat. Acad. Sci. U.S.A. 55 (1966), 240-243.

[6] On homotopy K 3 surfaces. Essays on Topology and Related Topics. Memoires dédies à G. de Rham. (1970) 58-69, Berlin-Heidelberg-New York : Springer.

[7] A certain type of irregular algebraic surfaces. J. d'Analyse Math., 19 (1967), 207-215.

[8] Holomorphic mappings of polydiscs into compact complex manifolds. J. Diff. Geometry, 6 (1971), 33-46.

[9] K. Kodaira : Collected Works. Tokyo : Iwanami, Princeton,N.J.: Princeton Univ. Press 1975.

Kodaira, K. and J. Morrow.

[1] Complex Manifolds. New York : Holt, Rinehalt and Winston, 1971.

Kodaira, K. and D. C. Spencer.

[1] On deformations of complex analytic structures, I and II. Ann. of Math., 67 (1958), 328-466.

[2] On deformations of complex analytic structures III. Stability theorms for complex structures. Ann. of Math., 71 (1960), 43-76.

Kuga, M.

[1] Fibre varieties over a symmetric space whos fibres are abelian varieties, I and II. Lecture note. Univ. of Chicago. 1963, 1964.

Kuhlmann, N.

[1] Projektive Modifikationen komplexer Räume. Math. Ann., 139 (1960), 217-238.

Lang, S.

[1] Introduction to Algebraic Geometry. New York;Interscience 1958.

Lejeune-Jalabert, M. and B. Teissier.

[1] Quelques calculs utiles pour la résolution des singularités. Mimeographed note, Ecole Polytechnique, 1972.

Maeda, H.

[1] Some complex structures on the product of spheres. J. Fac. Sci. Univ. Tokyo, Sec. I.A. 21(1974), 161-165.

Martynov, B. V.

[1] Three-dimensional varieties with rational sections. Math. Sbornik, 81 (123) (1970), 622-633, English translation, Math. U.S.S.R. Sbornik, 10 (1970), 569-579.

[2] An addendum to the article "Three-dimensional varieties with rational sections". Math. Sbornik 85 (127), (1971), 157-158, English translation, Math. U.S.S.R. Sbornik, 11 (1971), 157-158.

Maruyama, M.

[1] On classification of ruled surfaces. Lectures in Mathematics, Department of Kyoto Univ. 3, Tokyo : Kinokuniya 1970.

[2] On a family of algebraic vector bundles. Number Theory, Algebraic Geometry and Commutative Algebra, in honor of Y. Akizuki, 1973, 95-146, Kinokuniya.

Mather, J.

[1] Notes on topological stability. Mimeographed note. Harvard Univ. 1970.

Matsumura, H.

[1] On algebraic groups of birational transformations. Rend. Acad. Naz. Lincei, Ser. VII, 34 (1963), 151-155.

Matsumura, H. and F. Oort.

[1] Representability of group functors and automorphisms of algebraic schemes. Invent. Math., 4 (1967), 1-25.

Matsusaka, T.

[1] On canonically polarized varieties II. Amer. J. Math., 92 (1970), 283-292.

Matsushima, Y.

[1] Differentiable manifolds. New York : Marcel Dekker Inc. 1972.

[2] On Hodge manifolds with zero first Chern class. J. Diff. Geometry, 3 (1969), 477-480.

[3] Holomorphic Vector Fields on Compact Kähler Manifolds. Conf. Board Math. Sci. Regional Conf. Ser. in Math., 7, Amer. Math. Soc. 1971.

Milnor, J.

[1] Singular points of complex hypersurfaces. Ann. of Math. Studies, 61, Princeton Univ. Press 1968.

Miyaoka, Y.

[1] Kähler metrics on elliptic surfaces. Master detree thesis, Univ. of Tokyo 1974.

[2] Extension theorems of Kähler metrics. To appear in Proc. Japan Acad.

[3] Kähler metrics on elliptic surfaces. To appear in Proc. Japan Acad.

Moishezon, B.G.

[1] On n-dimensional compact varieties with n algebrically independent meromcrphic functions, I, II and III. Izv. Akad. Nauk SSSR Ser. Mat., 30 (1966), 133-174, 345-386, 621-656, English translation, AMS Translation Ser. 2, 63, 51-177.

Morita, S.

[1] A topological classification of complex structures on $S^1 \times \Sigma^{2n-1}$. To appear in Topology.

Mumford, D.

[1] Introduction to Algebraic Geometry. Cambridg, Mass.: Harvard Univ. Press.

[2] Lectures on Curves on an Algebraic Surface. Ann. of Math. Studies, 59, Princeton Univ. Press, 1966.

[3] Abelian Varieties. London: Oxford Univ. Press, 1970.

[4] The canonical ring of an algebraic surface. Appendix to Zariski's paper "The theorem of Riemann-Roch for high multiples of an effective divisors on an algebraic surface". Ann. of Math., 76 (1962), 612-615.

[5] Geometric Invariant Theory. Berlin-Heidelberg-New York : Springer 1965.

[6] An analytic construction of degenerating abelian varieties over complete rings. Compositio Math., 24 (1972), 239-272.

Murre, J. P.

[1] Reduction of the proof of the non-rationality of a non-singular cubic threefold to a result of Mumford. Compositio Math., 27 (1973), 63-82.

Nagata, M.

[1] On rational surfaces, I and II. Mem. Coll. Sci. Univ. Kyoto, 32 (1960), 351-370, ibid. 33 (1961), 271-293.

Nakai, Y.

[1] A criterion of an ample sheaf on a projective scheme. Amer. J. Math., 85 (1963), 14-26.

Nakamura, I.

[1] On classification of parallelisable manifolds and small deformations. To appear in J. Diff. Geometry.

[2] On degeneration of polarized abelian varieties. To appear.

Nakamura, I. and K. Ueno.

[1] An addition formula for Kodaira dimensions of analytic fibre

bundles whose fibre are Moišezon manifolds. J. Math. Soc. Japan, 25 (1973), 363-371.

Nakano, S.

[1] On the inverse of monoidal transformation. Publ. Research Inst. Math. Sci. Kyoto Univ. 6 (1971), 483-502.

Namikawa, Y.

[1] On the canonical holomorphic map from the moduli space of stable curves to the Igusa monoidal transform. Nagoya Math. J., 52 (1973), 197-259.

[2] Studies of degenerations. Classification of algebraic varieties and compact complex manifolds. Proceedings 1974. Lecture Notes in Math., 412, 165-210, Berlin-Heidelberg-New York: Springer 1974.

Namikawa, Y. and K. Ueno.

[1] The complete classification of fibres in pencils of curves of genus two. Manuscripta Math., 9 (1973), 163-186.

[2] On fibres in families of curves of genus two, I and II. Number theory, Algebraic Geometry and Commutative Algebra, in honor of Y. Akizuki, (1973), 297-371, Tokyo : Kinokuniya, to appear.

Narasimhan, R.

[1] Introduction to the theory of analytic spaces. Lecture Notes in Math., 25 . Berlin-Heidelberg-New York, Springer 1966.

Narasimhan, M. S. and R. R. Simha.

[1] Manifolds with ample canonical class. Invent. Math., 5 (1968), 120-128.

Oka, M.

[1] On the cohomology structure of projective varieties. Proc. Int. Conf. on Manifolds and Related Topics in Topology.137-144, Tokyo : Univ. of Tokyo Press 1974.

Oda, T. and K. Miyake.

[1] Almost homogeneous algebraic varieties under algebraic torus action. Proc. Int. Conf. on Manifolds and Related Topics in Topology 373-382, Tokyo : Univ. of Tokyo Press, 1974.

Potters, J.

[1] On almost homogeneous compact analytic surfaces. Invent. Math., 8 (1969) 244-266.

Popp, H.

[1] On moduli of algebraic varieties. I, II and III. Invent. Math. 22 (1973), 1-40, Compositio Math. 28 (1974), 51-81, to appear.

[2] Modulräume algebraischer Mannigfaltigkeiten. Classification of algebraic varieties and compact complex manifolds. Proceedings 1974. Lecture Notes in Math., 412, 219-242, Berlin-Heidelberg-New York : Springer 1974.

Pjateckiǐ-Šapiro, I. I. and I. R. Šafarevič.

[1] A Torelli theorem for algebraic surfaces of type K 3. Izv. Akad. Nauk SSSR Ser. Math., 35 (1971), 530-572, English translation, Math. USSR-Izvestija, 5 (1971), 547-588.

Remmert, R.

[1] Meromorphe Funktionen in Kompakten komplexen Räume. Math. Ann., 132 (1956), 277-288.

[2] Holomorphe und meromorphe Abbildungen komplexer Räume. Math. Ann., 133 (1957), 328-360.

Rimenschneider, O.

[1] Über die Anwendung algebraischer Methoden in der Deformationstheorie komplexer Räume. Math. Ann., 187 (1970), 40-55.

Rosenlicht, M.

[1] Some basic theorems on algebraic groups. Amer. J. Math., 78 (1956), 401-443.

Roth. L.

[1] Algebraic Threefolds (with special regard to problem of rationality). Berlin-Heidelberg-New York : Springer 1955.

Šafarevič, I. R. et al.

[1] Algebraic Surfaces. Proc. Steklov Institut. Moscwa, 1965. English translation. Providence, Rhode Island : Amer. Math. Soc., 1967.

Sakai, F.

[1] Degeneracy of holomorphic maps with ramification. To appear in Invent. Math.

Schlessinger, M.

[1] Regidity of quotient singularities. Invent. Math., 14 (1971), 17-26.

Serre, J. P.

[1] Un théorème de dualité. Comment. Math. Helvetia., 29 (1955), 9-26.

[2] Faisceaux algébriques cohérents. Ann. of Math., 61 (1955), 197-278.

[3] Géomètrie algébrique et géomètrie analytique. Ann. Inst. Fourier, 6 (1956), 1-42.

Siegel, C. L.

[1] Analytic functions of several complex variables. Mimeographed note, Princeton Univ. 1962, reprinted by Tokyo Univ. Press, 1970.

Spanier, E.

[1] The homology of Kummer manifolds. Proc. Amer. Math. Soc., 7 (1956), 155-160.

Stein, K.

[1] Meromorphic mappings. L'Ens. Math., Ser. II, 14 (1968), 29-46.

Suwa, T.

[1] On hyperelliptic surfaces. J. Fac. Sci. Univ. Tokyo. Sec. I, 14 (1970), 469-476.

[2] On ruled surfaces of genus 1. J. Math. Soc. Japan, 21 (1969), 291-311.

Thimm, W.

[1] Meromorphe Abbildungen von Riemannschen Bereichen. Math. Zeit. 60 (1954), 435-457.

Tjurin, A. N.

[1] Five lectures on three-dimensional varieties. Russian Math. Surveys, 27, No.3 (1972), 1-53.

Ueno, K.

[1] On fibre spaces of normally polarized abelian varieties of dimension 2. I, II and III. J. Fac. Sci. Univ. Tokyo, Sec. I. A., 17 (1971), 37-95, ibid. 19 (1972), 163-199, to appear.

[2] On Kodaira dimensions of certain algebraic varieties. Proc. Japan Acad., 47 (1971), 157-159.

[3] Classification of algebraic varieties. I, II. Compositio Math., 27 (1973), 277-342, to appear.

[4] Classification of algebraic varieties. Proc. Int. Conf. on Manifolds and Related topics in Topology. 329-332, Tokyo : Univ. of Tokyo Press 1974.

[5] Introduction to classification theory of algebraic varieties and compact complex spaces. Classification of algebraic varieties and compact complex manifolds. Proceedings 1974. Lecture Notes in Math., 412, 288-332, Berlin-Heidelberg-New York : Springer 1974.

[6] Introduction to classification theory of algebraic surfaces. Lecture note, Univ. of Amsterdam 1974.

[7] On the pluricanonical systems of algebraic manifolds of dimension 3. To appear.

[8] Algebraic surfaces which have pencils of curves of genus 2. To appear.

Wallace, A. H.

[1] Homology Theory of Algebraic Varieties, Oxford-London-New York-Braunschweig : Pergamon Press 1958.

Wang, H. C.

[1] Complex parallelisable manifolds. Proc. Amer. Math., Soc., 5 (1954), 771-776.

Weil, A.

[1] Foundation of Algebraic Geometry. Amer. Math. Soc. Colloquium Publ., 29 (revised edition) 1962.

[2] Introduction a L'Étude de Variétés Kähleriennes. Paris : Hermann, 1958.

Weyl, H.

[1] Die Idee der Riemannsche Flächen, revised edition. Stuttgard: Teubner 1955.

Zariski, O.

[1] Foundation of general theory of birational correspondences. Trans. A.M.S., 53 (1943), 490-542.

[2] Theory and applications of holomorphic functions on algebraic varieties over arbitrary ground field. Mem. Amer. Math. Soc., 5, 1951.

[3] Complete linear systems on normal varieties and a generalization of a lemma of Enriques-Severi. Ann. of Math., 55 (1952), 552-592.

[4] Introduction to the Problem of Minimal Models in the Theory of Algebraic Surfaces. Publ. Math. Soc. Japan, 4 (1958).

[5] The theorem of Riemann-Roch for high multiples of an effective divisor on an algebraic surface. Ann. of Math., 76 (1962), 560-615.

[6] Algebraic Surfaces, second edition. Berlin-Heidelberg-New York:

Springer, 1971.

[7] Oscar Zariski : Collected Papers, I, II. Cambridg, Mass.-London: The MIT Press 1972, 1973.

Zariski, O. and P. Samuel

[1] Commutative Algebra, vol. II. New York-London-Melbourne-New Delhi-Toronto : Van Nostrand, 1960.

Index

Vol. 342: Algebraic K-Theory II, "Classical" Algebraic K-Theory, and Connections with Arithmetic. Edited by H. Bass. XV, 527 pages. 1973. DM 36,–

Vol. 343: Algebraic K-Theory III, Hermitian K-Theory and Geometric Applications. Edited by H. Bass. XV, 572 pages. 1973. DM 38,–

Vol. 344: A. S. Troelstra (Editor), Metamathematical Investigation of Intuitionistic Arithmetic and Analysis. XVII, 485 pages. 1973. DM 34,–

Vol. 345: Proceedings of a Conference on Operator Theory. Edited by P. A. Fillmore. VI, 228 pages. 1973. DM 20,–

Vol. 346: Fučík et al., Spectral Analysis of Nonlinear Operators. II, 287 pages. 1973. DM 26,–

Vol. 347: J. M. Boardman and R. M. Vogt, Homotopy Invariant Algebraic Structures on Topological Spaces. X, 257 pages. 1973. DM 22,–

Vol. 348: A. M. Mathai and R. K. Saxena, Generalized Hypergeometric Functions with Applications in Statistics and Physical Sciences. VII, 314 pages. 1973. DM 26,–

Vol. 349: Modular Functions of One Variable II. Edited by W. Kuyk and P. Deligne. V, 598 pages. 1973. DM 38,–

Vol. 350: Modular Functions of One Variable III. Edited by W. Kuyk and J.-P. Serre. V, 350 pages. 1973. DM 26,–

Vol. 351: H. Tachikawa, Quasi-Frobenius Rings and Generalizations. XI, 172 pages. 1973. DM 18,–

Vol. 352: J. D. Fay, Theta Functions on Riemann Surfaces. V, 137 pages. 1973. DM 16,–

Vol. 353: Proceedings of the Conference on Orders, Group Rings and Related Topics. Organized by J. S. Hsia, M. L. Madan and T. G. Ralley. X, 224 pages. 1973. DM 20,–

Vol. 354: K. J. Devlin, Aspects of Constructibility. XII, 240 pages. 1973. DM 22,–

Vol. 355: M. Sion, A Theory of Semigroup Valued Measures. V, 140 pages. 1973. DM 16,–

Vol. 356: W. L. J. van der Kallen, Infinitesimally Central-Extensions of Chevalley Groups. VII, 147 pages. 1973. DM 16,–

Vol. 357: W. Borho, P. Gabriel und R. Rentschler, Primideale in Einhüllenden auflösbarer Lie-Algebren. V, 182 Seiten. 1973. DM 18,–

Vol. 358: F. L. Williams, Tensor Products of Principal Series Representations. VI, 132 pages. 1973. DM 16,–

Vol. 359: U. Stammbach, Homology in Group Theory. VIII, 183 pages. 1973. DM 18,–

Vol. 360: W. J. Padgett and R. L. Taylor, Laws of Large Numbers for Normed Linear Spaces and Certain Fréchet Spaces. VI, 111 pages. 1973. DM 16,–

Vol. 361: J. W. Schutz, Foundations of Special Relativity: Kinematic Axioms for Minkowski Space Time. XX, 314 pages. 1973. DM 26,–

Vol. 362: Proceedings of the Conference on Numerical Solution of Ordinary Differential Equations. Edited by D. Bettis. VIII, 490 pages. 1974. DM 34,–

Vol. 363: Conference on the Numerical Solution of Differential Equations. Edited by G. A. Watson. IX, 221 pages. 1974. DM 20,–

Vol. 364: Proceedings on Infinite Dimensional Holomorphy. Edited by T. L. Hayden and T. J. Suffridge. VII, 212 pages. 1974. DM 20,–

Vol. 365: R. P. Gilbert, Constructive Methods for Elliptic Equations. VII, 397 pages. 1974. DM 26,–

Vol. 366: R. Steinberg, Conjugacy Classes in Algebraic Groups (Notes by V. V. Deodhar). VI, 159 pages. 1974. DM 18,–

Vol. 367: K. Langmann und W. Lütkebohmert, Cousinverteilungen und Fortsetzungssätze. VI, 151 Seiten. 1974. DM 16,–

Vol. 368: R. J. Milgram, Unstable Homotopy from the Stable Point of View. V, 109 pages. 1974. DM 16,–

Vol. 369: Victoria Symposium on Nonstandard Analysis. Edited by A. Hurd and P. Loeb. XVIII, 339 pages. 1974. DM 26,–

Vol. 370: B. Mazur and W. Messing, Universal Extensions and One Dimensional Crystalline Cohomology. VII, 134 pages. 1974. DM 16,–

Vol. 371: V. Poenaru, Analyse Différentielle. V, 228 pages. 1974. DM 20,–

Vol. 372: Proceedings of the Second International Conference on the Theory of Groups 1973. Edited by M. F. Newman. VII, 740 pages. 1974. DM 48,–

Vol. 373: A. E. R. Woodcock and T. Poston, A Geometrical Study of the Elementary Catastrophes. V, 257 pages. 1974. DM 22,–

Vol. 374: S. Yamamuro, Differential Calculus in Topological Linear Spaces. IV, 179 pages. 1974. DM 18,–

Vol. 375: Topology Conference 1973. Edited by R. F. Dickman Jr. and P. Fletcher. X, 283 pages. 1974. DM 24,–

Vol. 376: D. B. Osteyee and I. J. Good, Information, Weight of Evidence, the Singularity between Probability Measures and Signal Detection. XI, 156 pages. 1974. DM 16,–

Vol. 377: A. M. Fink, Almost Periodic Differential Equations. VIII, 336 pages. 1974. DM 26,–

Vol. 378: TOPO 72 – General Topology and its Applications. Proceedings 1972. Edited by R. Alò, R. W. Heath and J. Nagata. XIV, 651 pages. 1974. DM 50,–

Vol. 379: A. Badrikian et S. Chevet, Mesures Cylindriques, Espaces de Wiener et Fonctions Aléatoires Gaussiennes. X, 383 pages. 1974. DM 32,–

Vol. 380: M. Petrich, Rings and Semigroups. VIII, 182 pages. 1974. DM 18,–

Vol. 381: Séminaire de Probabilités VIII. Edité par P. A. Meyer. IX, 354 pages. 1974. DM 32,–

Vol. 382: J. H. van Lint, Combinatorial Theory Seminar Eindhoven University of Technology. VI, 131 pages. 1974. DM 18,–

Vol. 383: Séminaire Bourbaki – vol. 1972/73. Exposés 418-435. IV, 334 pages. 1974. DM 30,–

Vol. 384: Functional Analysis and Applications, Proceedings 1972. Edited by L. Nachbin. V, 270 pages. 1974. DM 22,–

Vol. 385: J. Douglas Jr. and T. Dupont, Collocation Methods for Parabolic Equations in a Single Space Variable (Based on C¹-Piecewise-Polynomial Spaces). V, 147 pages. 1974. DM 16,–

Vol. 386: J. Tits, Buildings of Spherical Type and Finite BN-Pairs. IX, 299 pages. 1974. DM 24,–

Vol. 387: C. P. Bruter, Eléments de la Théorie des Matroïdes. V, 138 pages. 1974. DM 18,–

Vol. 388: R. L. Lipsman, Group Representations. X, 166 pages. 1974. DM 20,–

Vol. 389: M.-A. Knus et M. Ojanguren, Théorie de la Descente et Algèbres d' Azumaya. IV, 163 pages. 1974. DM 20,–

Vol. 390: P. A. Meyer, P. Priouret et F. Spitzer, Ecole d'Eté de Probabilités de Saint-Flour III – 1973. Edité par A. Badrikian et P.-L. Hennequin. VIII, 189 pages. 1974. DM 20,–

Vol. 391: J. Gray, Formal Category Theory: Adjointness for 2-Categories. XII, 282 pages. 1974. DM 24,–

Vol. 392: Géométrie Différentielle, Colloque, Santiago de Compostela, Espagne 1972. Edité par E. Vidal. VI, 225 pages. 1974. DM 20,–

Vol. 393: G. Wassermann, Stability of Unfoldings. IX, 164 pages. 1974. DM 20,–

Vol. 394: W. M. Patterson 3rd, Iterative Methods for the Solution of a Linear Operator Equation in Hilbert Space – A Survey. III, 183 pages. 1974. DM 20,–

Vol. 395: Numerische Behandlung nichtlinearer Integrodifferential- und Differentialgleichungen. Tagung 1973. Herausgegeben von R. Ansorge und W. Törnig. VII, 313 Seiten. 1974. DM 28,–

Vol. 396: K. H. Hofmann, M. Mislove and A. Stralka, The Pontryagin Duality of Compact O-Dimensional Semilattices and its Applications. XVI, 122 pages. 1974. DM 18,–

Vol. 397: T. Yamada, The Schur Subgroup of the Brauer Group. V, 159 pages. 1974. DM 18,–

Vol. 398: Théories de l'Information, Actes des Rencontres de Marseille-Luminy, 1973. Edité par J. Kampé de Fériet et C. Picard. XII, 201 pages. 1974. DM 23,–

Vol. 277: Séminaire Banach. Edité par C. Houzel. VII, 229 pages. 1972. DM 20,–

Vol. 278: H. Jacquet, Automorphic Forms on GL(2). Part II. XIII, 142 pages. 1972. DM 16,–

Vol. 279: R. Bott, S. Gitler and I. M. James, Lectures on Algebraic and Differential Topology. V, 174 pages. 1972. DM 18,–

Vol. 280: Conference on the Theory of Ordinary and Partial Differential Equations. Edited by W. N. Everitt and B. D. Sleeman. XV, 367 pages. 1972. DM 26,–

Vol. 281: Coherence in Categories. Edited by S. Mac Lane. VII, 235 pages. 1972. DM 20,–

Vol. 282: W. Klingenberg und P. Flaschel, Riemannsche Hilbertmannigfaltigkeiten. Periodische Geodätische. VII, 211 Seiten. 1972. DM 20,–

Vol. 283: L. Illusie, Complexe Cotangent et Déformations II. VII, 304 pages. 1972. DM 24,–

Vol. 284: P. A. Meyer, Martingales and Stochastic Integrals I. VI, 89 pages. 1972. DM 16,–

Vol. 285: P. de la Harpe, Classical Banach-Lie Algebras and Banach-Lie Groups of Operators in Hilbert Space. III, 160 pages. 1972. DM 16,–

Vol. 286: S. Murakami, On Automorphisms of Siegel Domains. V, 95 pages. 1972. DM 16,–

Vol. 287: Hyperfunctions and Pseudo-Differential Equations. Edited by H. Komatsu. VII, 529 pages. 1973. DM 36,–

Vol. 288: Groupes de Monodromie en Géométrie Algébrique. (SGA 7 I). Dirigé par A. Grothendieck. IX, 523 pages. 1972. DM 50,–

Vol. 289: B. Fuglede, Finely Harmonic Functions. III, 188. 1972. DM 18,–

Vol. 290: D. B. Zagier, Equivariant Pontrjagin Classes and Applications to Orbit Spaces. IX, 130 pages. 1972. DM 16,–

Vol. 291: P. Orlik, Seifert Manifolds. VIII, 155 pages. 1972. DM 16,–

Vol. 292: W. D. Wallis, A. P. Street and J. S. Wallis, Combinatorics: Room Squares, Sum-Free Sets, Hadamard Matrices. V, 508 pages. 1972. DM 50,–

Vol. 293: R. A. DeVore, The Approximation of Continuous Functions by Positive Linear Operators. VIII, 289 pages. 1972. DM 24,–

Vol. 294: Stability of Stochastic Dynamical Systems. Edited by R. F. Curtain. IX, 332 pages. 1972. DM 26,–

Vol. 295: C. Dellacherie, Ensembles Analytiques, Capacités, Mesures de Hausdorff. XII, 123 pages. 1972. DM 16,–

Vol. 296: Probability and Information Theory II. Edited by M. Behara, K. Krickeberg and J. Wolfowitz. V, 223 pages. 1973. DM 20,–

Vol. 297: J. Garnett, Analytic Capacity and Measure. IV, 138 pages. 1972. DM 16,–

Vol. 298: Proceedings of the Second Conference on Compact Transformation Groups. Part 1. XIII, 453 pages. 1972. DM 32,–

Vol. 299: Proceedings of the Second Conference on Compact Transformation Groups. Part 2. XIV, 327 pages. 1972. DM 26,–

Vol. 300: P. Eymard, Moyennes Invariantes et Représentations Unitaires. II. 113 pages. 1972. DM 16,–

Vol. 301: F. Pittnauer, Vorlesungen über asymptotische Reihen. VI, 186 Seiten. 1972. DM 18,–

Vol. 302: M. Demazure, Lectures on p-Divisible Groups. V, 98 pages. 1972. DM 16,–

Vol. 303: Graph Theory and Applications. Edited by Y. Alavi, D. R. Lick and A. T. White. IX, 329 pages. 1972. DM 26,–

Vol. 304: A. K. Bousfield and D. M. Kan, Homotopy Limits, Completions and Localizations. V, 348 pages. 1972. DM 26,–

Vol. 305: Théorie des Topos et Cohomologie Etale des Schémas. Tome 3. (SGA 4). Dirigé par M. Artin, A. Grothendieck et J. L. Verdier. VI, 640 pages. 1973. DM 50,–

Vol. 306: H. Luckhardt, Extensional Gödel Functional Interpretation. VI, 161 pages. 1973. DM 18,–

Vol. 307: J. L. Bretagnolle, S. D. Chatterji et P.-A. Meyer, Ecole d'été de Probabilités: Processus Stochastiques. VI, 198 pages. 1973. DM 20,–

Vol. 308: D. Knutson, λ-Rings and the Representation Theory of the Symmetric Group. IV, 203 pages. 1973. DM 20,–

Vol. 309: D. H. Sattinger, Topics in Stability and Bifurcation Theory. VI, 190 pages. 1973. DM 18,–

Vol. 310: B. Iversen, Generic Local Structure of the Morphisms in Commutative Algebra. IV, 108 pages. 1973. DM 16,–

Vol. 311: Conference on Commutative Algebra. Edited by J. W. Brewer and E. A. Rutter. VII, 251 pages. 1973. DM 22,–

Vol. 312: Symposium on Ordinary Differential Equations. Edited by W. A. Harris, Jr. and Y. Sibuya. VIII, 204 pages. 1973. DM 22,–

Vol. 313: K. Jörgens and J. Weidmann, Spectral Properties of Hamiltonian Operators. III, 140 pages. 1973. DM 16,–

Vol. 314: M. Deuring, Lectures on the Theory of Algebraic Functions of One Variable. VI, 151 pages. 1973. DM 16,–

Vol. 315: K. Bichteler, Integration Theory (with Special Attention to Vector Measures). VI, 357 pages. 1973. DM 26,–

Vol. 316: Symposium on Non-Well-Posed Problems and Logarithmic Convexity. Edited by R. J. Knops. V, 176 pages. 1973. DM 18,–

Vol. 317: Séminaire Bourbaki – vol. 1971/72. Exposés 400–417. IV, 361 pages. 1973. DM 26,–

Vol. 318: Recent Advances in Topological Dynamics. Edited by A. Beck, VIII, 285 pages. 1973. DM 24,–

Vol. 319: Conference on Group Theory. Edited by R. W. Gatterdam and K. W. Weston. V, 188 pages. 1973. DM 18,–

Vol. 320: Modular Functions of One Variable I. Edited by W. Kuyk. V, 195 pages. 1973. DM 18,–

Vol. 321: Séminaire de Probabilités VII. Edité par P. A. Meyer. VI, 322 pages. 1973. DM 26,–

Vol. 322: Nonlinear Problems in the Physical Sciences and Biology. Edited by I. Stakgold, D. D. Joseph and D. H. Sattinger. VIII, 357 pages. 1973. DM 26,–

Vol. 323: J. L. Lions, Perturbations Singulières dans les Problèmes aux Limites et en Contrôle Optimal. XII, 645 pages. 1973. DM 42,–

Vol. 324: K. Kreith, Oscillation Theory. VI, 109 pages. 1973. DM 16,–

Vol. 325: Ch.-Ch. Chou, La Transformation de Fourier Complexe et L'Equation de Convolution. IX, 137 pages. 1973. DM 16,–

Vol. 326: A. Robert, Elliptic Curves. VIII, 264 pages. 1973. DM 22,–

Vol. 327: E. Matlis, 1-Dimensional Cohen-Macaulay Rings. XII, 157 pages. 1973. DM 18,–

Vol. 328: J. R. Büchi and D. Siefkes, The Monadic Second Order Theory of All Countable Ordinals. VI, 217 pages. 1973. DM 20,–

Vol. 329: W. Trebels, Multipliers for (C, α)-Bounded Fourier Expansions in Banach Spaces and Approximation Theory. VII, 103 pages. 1973. DM 16,–

Vol. 330: Proceedings of the Second Japan-USSR Symposium on Probability Theory. Edited by G. Maruyama and Yu. V. Prokhorov. VI, 550 pages. 1973. DM 36,–

Vol. 331: Summer School on Topological Vector Spaces. Edited by L. Waelbroeck. VI, 226 pages. 1973. DM 20,–

Vol. 332: Séminaire Pierre Lelong (Analyse) Année 1971-1972. V, 13[illegible] pages. 1973. DM 16,–

Vol. 333: Numerische, insbesondere approximationstheoretische Behandlung von Funktionalgleichungen. Herausgegeben von R. Ansorge und W. Törnig. VI, 296 Seiten. 1973. DM 24,–

Vol. 334: F. Schweiger, The Metrical Theory of Jacobi-Perron Algorithm. V, 111 pages. 1973. DM 16,–

Vol. 335: H. Huck, R. Roitzsch, U. Simon, W. Vortisch, R. Walden, B. Wegner und W. Wendland, Beweismethoden der Differentialgeometrie im Großen. IX, 159 Seiten. 1973. DM 18,–

Vol. 336: L'Analyse Harmonique dans le Domaine Complexe. Edité par E. J. Akutowicz. VIII, 169 pages. 1973. DM 18,–

Vol. 337: Cambridge Summer School in Mathematical Logic. Edited by A. R. D. Mathias and H. Rogers. IX, 660 pages. 1973. DM 42,–

Vol. 338: J. Lindenstrauss and L. Tzafriri, Classical Banach Spaces. IX, 243 pages. 1973. DM 22,–

Vol. 339: G. Kempf, F. Knudsen, D. Mumford and B. Saint-Donat, Toroidal Embeddings I. VIII, 209 pages. 1973. DM 20,–

Vol. 340: Groupes de Monodromie en Géométrie Algébrique. (SGA 7 II). Par P. Deligne et N. Katz. X, 438 pages. 1973. DM 40,–

Vol. 341: Algebraic K-Theory I, Higher K-Theories. Edited by H. Bass. XV, 335 pages. 1973. DM 26,–